Torill Kornfeldt

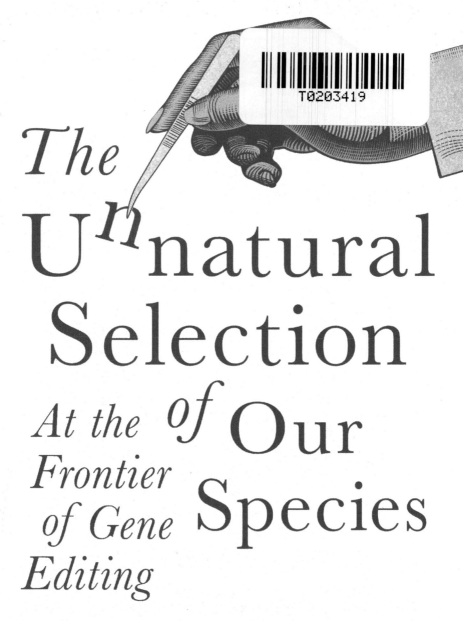

The
U*n*natural
Selection
At the of Our
Frontier
of Gene Species
Editing

Translated by Fiona Graham

Hero, an imprint of
Legend Times Group Ltd, 51 Gower Street, London, WC1E 6HJ
hero@hero-press.com | www.hero-press.com

© Torill Kornfeldt 2021
The right of the above author to be identified as the author of this
work has been asserted in accordance with the Copyright, Designs
and Patents Act 1988. British Library Cataloguing in Publication
Data available.

Translation © Fiona Graham 2021

First published in Swedish in 2020 by Natur & Kultur
Published by agreement with the Kontext Agency

The cost of this translation was defrayed by a subsidy from the
Swedish Arts Council, gratefully acknowledged

Illustrations: Erik Kohlström
Cover design: Sophie Burdess

Print ISBN: 978-1-91564-353-7
Ebook ISBN: 978-1-80031-343-9
Set in Times. Printing managed by Jellyfish Solutions Ltd

Torill Kornfeldt is a Swedish science journalist with a background in biology. She has worked in the science department of Sweden's leading morning newspaper Dagens Nyheter and at the science branch of the Swedish public service radio. There she created the successful radio show Tekniksafari (Tech Safari) on new technology changing society. Her main focus is on how emerging bioengineering and technology will shape our future.

Fiona Graham has a degree in Modern Languages from Oxford University, and has lived in Kenya, Germany, the Netherlands, Luxembourg, Nicaragua and Belgium. She translates from Spanish, French, Dutch, Swedish and German, and is currently the reviews editor at the Swedish Book Review.

Contents

Introduction

A Time of Radical Change

The 2020 Nobel Prize festivities broke with tradition.[1] No banquet in the Blue Hall or dancing in the Golden Hall, no ceremonial presentation of medals by the King of Sweden. Instead, the laureates received their medals and congratulations individually, in many cases from Swedish ambassadors around the world. The festivities are on hold until the pandemic has subsided.

Despite the exceptional circumstances, the two chemistry laureates attracted rather more attention than usual. A Nobel Prize is rarely awarded just a few years after the groundbreaking discovery it honours; in most cases, decades pass before important discoveries have proven their value and 'conferred the greatest benefit to humankind'. Even though the results for which the prize was awarded were only published in 2012, it is already clear that the laureates' creation is poised to change the world – and us – in fundamental ways.

The discovery made by these two scientists, Jennifer Doudna and Emmanuelle Charpentier, is actually quite simple: a chemical device developed by bacteria can be turned into a tool that we humans can use. Yet that tool – the 'gene scissors' CRISPR-Cas9 – has transformed our potential future.

The new generation of sophisticated, precise and cheap gene technology has opened up infinite opportunities in medicine, opportunities to cure people and improve their lives. Yet these also involve new risks, as well as concerns about how best to use the technology. And this isn't just about the future: new though the technology is, it has already had a major impact. That was why the Nobel Committee decided to award the prize to Doudna and Charpentier as early as 2020.

During the ten years and more that I've been working in science journalism, I have witnessed rapid advances in what new research enables us to achieve. We now have the capacity to redraw the blueprint of our own species, along with that of all the animals and plants in the world. We can transform our own genetic makeup and design living beings that have never existed before. Some of these developments are described in my previous book, *The Re-Origin of Species*.

When I started writing about these scientific advances as a news reporter, I realized that the people I met outside my professional circle were still talking about gene technology and gene-editing as if they lay in the distant future, as if the phenomenon of gene-editing in humans were something that might confront their grandchildren's generation – if that. Most seemed to dismiss the idea as pure science fiction. Yet many of our wildest dreams and our worst nightmares are already scientific reality, or at least well on the way there. Very few people outside scientific circles seem to be aware of this. There is still little or no public debate on the risks and opportunities the new gene technology presents, although it's bound to change our society in many ways.

My aim in writing this book is to depict the rapid developments we are in the midst of and to show their possible consequences. I have tried to sketch these advances while they are happening. To do so, I have made every effort to meet the people in the vanguard of this new field of research. I have interviewed many scientists, as well as patients, activists and parents who want to change their children's genetic makeup. I have visited commercial clinics that provide gene therapy and seen laboratory animals that have been genetically modified in various ways.

To understand all of this, I needed to travel so that I could see the research and its results *in situ*. I spent nearly a month in China – which is on the way to becoming the world leader in gene technology – and took high-speed trains to different parts of the country. In Beijing I met scientists who had edited the DNA of human embryos; in leafy Hangzhou I talked to cancer patients and doctors experimenting with gene technology as a cure for cancer; and in Kunming, the city of eternal spring in China's south-west, I visited a centre for research on genetically modified monkeys. And there was much more. This journey also showed me clearly how China is flexing its scientific muscles and gearing up to overtake the United States and Europe. The US–Soviet space race has been replaced by a scientific arms race, a gene race. While in Asia, I also met scientists and visited clinics in Japan, South Korea and Hong Kong.

But though China and Asia in general are playing an increasingly important role in scientific progress, developments in this field are not confined to Asia. I visited biohackers in a New York lab who are part of a non-mainstream movement keen to share gene technology with the masses, attended scientific conferences and talked to activists in Boston about who actually gets access to the costly new drugs developed with the help of modern gene technology. I travelled to a fascinating research centre in the countryside just outside Chicago to meet a very special genetically modified pig. This book is a piece of reportage that presents a snapshot of where we are now.

But technology isn't merely a technical matter; it's not just about what scientists do in labs or operating theatres. Technical developments must also accommodate us – human beings, our society and our culture. And our view of both risks and opportunities is formed by our narratives; technical advances have to be fitted into our view of the world. Just over 200 years ago, an adolescent girl sat in a drawing room near Lake Geneva, writing what was to become one of the world's most famous books, a novel that paved the way for a whole new genre. That time, just like the present day, was an era of radical change in which rapid scientific advances were transforming the world.[2] The brilliant young lady was Mary Wollstonecraft Godwin, whose novel *Frankenstein* introduced a new way of looking at the possibility of transforming our own bodies.[3] Her story of the bold, curious scientist has become iconic.

Just like the scientist Victor Frankenstein, we face a choice: what do we do with the knowledge and the power we've acquired?

Like all science fiction, *Frankenstein* takes an idea and pursues it to its conclusion: what would happen if a scientist succeeded in creating life – if we humans had the same powers as God? The scientific breakthrough is just the starting point: most of the story concerns what happens afterwards. When Victor Frankenstein sees the result of his experiment, he is terrified and flees. He leaves his newborn creature all alone, inarticulate and confused. The creature clearly isn't evil from the outset – it's more like an ingenuous child – but Victor's actions drive it to murder. This is a tale that captures the zeitgeist, the feeling of living in a world where the ground moves under your feet and all established truths have to be re-examined.

The novel has been interpreted and reinterpreted countless times, but what struck me on a first reading was a sense of sadness that so great a scientific advance was wasted and ended in tragedy: of disappointment that Victor's knowledge fails to make the world better. For, despite his egoism and arrogance, I can't help but identify with him. As the old sayings have it, you shouldn't play with fire, and curiosity killed the cat – but I wouldn't have been able to resist experimenting if I'd been in his shoes. It's hard not to like someone who is prepared to defy the gods to give humankind more knowledge – whatever the consequences.

For two centuries, the tale of Victor Frankenstein, the reckless scientist, has shaped our view of what happens when human beings seek to control nature. In recent decades the narrative has increasingly focused on gene technology and scientists' ability to reprogramme the code within our cells, the code that controls life itself. Genetically modified crops are called 'Frankenfood', the first sheep ever cloned was compared to Frankenstein's monster, and the tale of the irresponsible scientist comes up for discussion each time the public thinks science has gone 'too far' or become 'unnatural'. Frankenstein has become a code word understood by everyone, regardless of whether they've read the book, seen one of the many films or worn a Halloween costume based on the Frankenstein story.

Yet there's another narrative that has shaped our view of science and its limitless potential, a heartfelt exhortation to 'boldly go where no man [or no one] has gone before'. That famous phrase is the signature of *Star Trek*, a space odyssey driven by the same ethos since its very first season in 1966.[4] Since then, we've been able to follow the adventures of the Starship *Enterprise*'s crew, through various films and series, for over half a century. Essentially, the stories depict a highly developed utopian society, based on the idea that advanced research and technology have not just created a better life for all, but have also raised human beings to a higher plane of development. Many of humankind's oldest afflictions – hunger, disease and want – have been eradicated. This is a society that has risen above the barbaric phenomena of the past, such as racism and sexism. It's clearly no coincidence that one of the first ever televised kisses between an African American and a white American happened in this series.[5] *Star Trek* is full of post-war optimism: the threats to humanity it depicts are totalitarianism and militarism, not science that has 'gone too far'. The series is underpinned by a strong fundamental sense of optimism about the future.

Of course, *Star Trek* wasn't the first fictional narrative to show humanity using knowledge to reach new heights. Similarly, the tale of Frankenstein's hubris and fall reflects hundreds of other stories.

It may seem a little odd to focus on classic science fiction in a book about the cutting edge of scientific research, but for good or ill we humans tend to think in terms of stories. The tales of adventure we've told each other throughout history lay the foundations for the way we envisage the future. They frame our perceptions of new scientific discoveries, providing us with a lens through which to understand the world.

The problem isn't that both *Star Trek* and *Frankenstein* are science fiction, it's that they're true. Our view of science has two main facets: on the one hand, anxiety that progress will be too rapid and have fatal consequences, and, on the other, hope that progress can free us from suffering and barbarity. Anxiety and hope together determine which experiments we choose to conduct, and what impact their results have on society.[6]

New science and technology have transformed and enhanced the lives we lead. We Westerners live in a world where people are far more prosperous, healthy and secure than ever before, a future almost fit for the Starship *Enterprise*.[7] Nearly all the good things in our everyday lives – waking up in a bed that's not stuffed with straw or crawling with bugs; having the option of refrigerated, pasteurized milk and yoghurt with sliced banana for breakfast; or walking home along well-lit streets in the dark of a November evening – are the result of research and technological development. Yet at the same time we live in a world facing new and intractable problems. We have all the knowledge we need to ensure that no one has to starve, yet the same technological advances have resulted in climate change, emissions of toxic chemicals and a global extinction crisis. It is our fault that more city-dwellers are now dying from air pollution than, for example, from diabetes or AIDS.[8] We have created an infinitely better world, and yet our very creations have turned against us, causing one problem after another.

These two narratives are not mutually exclusive: there's no need to choose between them. The challenge, rather, is to hold both of them in your mind at the same time: exuberant, unbridled optimism and profound fear – sometimes thoughtful, sometimes irrational. The rapid progress we're seeing in gene technology makes it a field in which the two narratives need to come together and unite if we are to be able to navigate the future. For gene technology, just like earlier major breakthroughs such as electricity or splitting the atom, has the potential to change our society fundamentally, for both good and ill. We have to find a way forward that allows us to remain open to its potential without succumbing to the worst of its concomitant pitfalls. In this book, I look at both the dangers of gene technology and its incredible potential. It is an eye-opening journey, for though I'm familiar with the achievements in this field, the events of November 2018 astounded me.

* * *

I had a plan for this book. I knew what issues I wanted to raise, and in what order: which gene technology scenarios were reasonable – credible – within the near future, and which would take longer to materialize. I thought developments could be summarized as follows: new methods for genetically modifying cells have huge potential to cure medical conditions and correct the things that go wrong with our bodies. They will revolutionize treatments for countless diseases, but will mostly involve tweaking a few cells in adults' bodies: repairing an eye, shrinking a cancer till it disappears, helping a damaged organ to function better. It will be a good many years until the first genetically modified baby is born. And in fact there was a consensus on that point among most scientists. When it happened, so they thought, it would be a calm, controlled process, the culmination of mature reflection within the scientific community.

But the world never ceases to amaze us.

On 26 November 2018 the Chinese scientist He Jiankui demolished all these notions about the future when he announced on YouTube that he was the first person in the world to have edited a gene in two embryos which had then been implanted in a woman's womb – and survived. The first genetically modified humans had been born.

'I understand that my work will be controversial, but I believe families need this technology, and I'm willing to take criticism,' he says in the video. He had genetically edited twin girls pseudonymously called Lulu and Nana, giving them a modification he hoped would shield them from infection with the HIV virus.[9] The day after, he stood before the world's TV cameras and hundreds of outraged scientists to give an account of the experiment at a conference in Hong Kong. He Jiankui is still a young man, just 34 years old, and it was clear how nervous he was when he walked onto the podium, but once he started to speak, his conviction that he had done the right thing shone through his nervous manner.

This was the moment that the world's geneticists had feared. Since the revolutionary gene technology CRISPR was developed just a few years ago, it had been clear that it could be used to alter human genes, both in adults and in embryos, but that was precisely why He Jiankui's experiment provoked such an uproar.

In my previous book I wrote about the attempts being made to recreate extinct animals, and the consequences which that might have.[10] The project grew into a book about the Siberian tundra, the possibility of miniature rhinos as pets, the cloning of dogs and the genetic modification of corals. But there was one topic I avoided throughout – the human body. That seemed too vast and frightening a subject. And yet it was unavoidable. For even if we use the new gene technology to alter everything around us – our food, our fuels, our pets, our clothes, our perfumes, the natural world – the greatest revolution will be that within ourselves. The idea for this book has buzzed around in my head like an irritating fly ever since. The major breakthroughs of recent years made up my mind for me: I had to write a book about the new gene technology.

But before we take a detailed look at these advances, or at what exactly genes are and how gene technology works, there are two names we need to bear in mind – Louise Brown and Jesse Gelsinger.

The birth of Louise Brown by Caesarean section in July 1978 was a scientific sensation.[11] This British baby was the first person ever born thanks to in-vitro fertilization (IVF). Her parents had been trying in vain for years to have children, but her mother had blocked fallopian tubes. A team of scientists decided to take eggs and sperm from the parents, combine them to form an embryo, and then try to implant the embryo in the mother's womb, where it was hoped it would develop as normal. This method was already being used in other animals, including cows, but had never been applied to humans. The team was led by Robert Edwards, who had been studying human fertility since the 1960s and had succeeded in getting eggs and sperm to combine in the laboratory. He worked together with Patrick Steptoe, a gynaecologist, and Jean Purdy, a nurse and embryologist. She was the first to see the cells that would eventually become Louise divide and grow.[12]

The trio had been trying unsuccessfully for years to develop a method of in-vitro fertilization and had met strong resistance. It was hard to get funding for their work, and they had been taken to court on several occasions. An article published in 1972 in the British magazine *Nova* described IVF as 'the biggest threat since the atom bomb' and called on the public to regulate research. The idea of producing children in this way was seen as both unnatural and immoral.[13]

But Louise wasn't to remain unique for very long. Just over two months later, a second baby girl was born in India using a different method of IVF developed by a scientist there. When Louise's little sister came into the world four years later, she was the fortieth baby to be born through IVF. In 2010, Robert Edwards, the only one of the original team still alive, was awarded the Nobel Prize in Medicine. To date, over 8 million children have been born worldwide through assisted reproductive technology, IVF and related methods.[14] Resistance to this method has melted away, and the fact that couples have recourse to such treatment if they can't or don't want to conceive in another way is seen as quite uncontroversial in most cases. Today Louise Brown is over 40, lives in Britain and has two children of her own.

Just three years after Louise, in June 1981, Jesse Gelsinger was born on the other side of the Atlantic.[15] There were no scientific fanfares, just worried parents and doctors. Jesse had a genetic disease that meant his body couldn't break down protein without poisoning him. A bite of hamburger or sausage was enough to put him in a coma. Most people with Jesse's disease (OTCD) die early in life. Only one in 80,000 survives their first five years. Jesse, however, had a relatively mild variant, so he was able to cope with the help of drugs and a strict diet. But as an 18-year-old in 1999 he heard about a new clinical trial to find a cure for the disease. He volunteered to take part.

The scientist leading the study, James Wilson, was a pioneer in the brand-new field of gene therapy, the aim of which was to repair flawed genetic material by inserting new, functioning genes into the body's cells. A few years previously, he and other scientists had succeeded in helping a woman with a genetic liver disorder, and most experts believed gene therapy was set to revolutionize medicine. Public funding and money from the dot-com era gave the research an extra boost, and by 1999 there were nearly a hundred clinical trials involving gene therapy in humans. It's reasonable to assume that Jesse Gelsinger was feeling cheerful and optimistic when he had his first jab on 13 September. But just four days later he was dead. His body had overreacted to the injection. The scientists had used a virus as the vector to deliver the new genes to his cells, but it made his immune system go haywire. His body swelled up, his liver stopped working, he developed a fever and finally he fell into a coma, after which it was impossible to save his life.

After Jesse's death, the clinical trial was called off and James Wilson and the other researchers involved were investigated. How could things have gone so wrong? It emerged that the scientists had made several mistakes. Wilson was banned for five years from research on human subjects. Yet Jesse's violent reaction to that particular treatment was mainly a matter of chance, ignorance and bad luck. Looking at the sad outcome with the benefit of hindsight, it doesn't seem so surprising. Different test subjects react differently, but in this case other participants also experienced serious side effects, and lab animals had died. It is now known that our immune system can react strongly to the type of virus that had been used. As a result of Jesse's death, not only were James Wilson's experiments stopped, but, in practice, the whole field of gene therapy died. There was no funding and no interest after the tragedy, and Jesse's death began to be used as an example of how badly wrong things can go when science tries to go 'too far'.

It would take over a decade for gene-therapy research to recover. It's now making a comeback, however, and observers again predict that it will revolutionize medicine. Today, other viruses are often used to deliver genetic material, or other delivery systems are used, to avoid overreaction by the immune system. The new wave of gene technology is closely linked with the technique used on the first gene-edited humans.

There are plenty of instances throughout history of scientists who have taken independent action and been judged harshly by their contemporaries, even when they have made invaluable discoveries. One well-known example is Ignaz Semmelweis.[16] He was a physician in a maternity department in Austria, where he realized after a while that women ran less risk of dying of post-partum infections if the doctors attending them simply washed their hands. That may seem obvious today, but when he published his research in 1847 it almost unleashed a scandal. How dare he suggest that gentlemen – and doctors were gentlemen – didn't have clean hands, but needed to wash them before operations? As if they were manual labourers! To make things worse, it was common practice at the time for a doctor to attend a birth just after conducting a post-mortem in the morgue, after which he would merely wipe his hands on a rag.

One of the problems was that Semmelweis had no explanation for why hand disinfection helped: it was not yet known that bacteria could spread disease. He succeeded in reducing mortality in his own hospital, but no other physicians followed suit. In the end he became depressed because no one would listen to his ideas. He was sent to a lunatic asylum against his will and died shortly afterwards. And the hospital where he had worked rejected his crazy ideas and abandoned the practice of hand disinfection, whereupon the mortality rate rose again.

Other scientists have been judged far more harshly by history than by their own time. One example is the experiments conducted at the Vipeholm institute for the mentally disabled in Lund, Sweden, from 1945 to 1955, a project that showed the link between caries and sugar by dispensing sweets to people with cognitive disabilities. These experiments, which went on for nearly ten years, completely ruined many of these people's teeth.[17]

There are also questions and fuzzy areas around the way in which the experiments involving Louise and Jesse were conducted. Were the participants really given all the information they needed to make a well-informed decision? The two cases had opposite outcomes: one experiment resulted in success and a Nobel Prize, the other in a tragedy. I think we need to remember both of them when we look at how gene technology could change the world. It's not at all clear in advance who's going to be a hero, and who's going to create a monster.

The incredibly rapid developments in gene technology over the last few years present us with many new choices. Who's to decide which cells and which bodies are to be modified? Who decides whether to prioritize a given disorder, or whether would-be parents who carry genes associated with serious diseases should be allowed to gene-edit their offspring? Should such decisions be taken by states or by individuals? If the former, which states, and if the latter, which individuals?

This is a book of reportage about how humanity's very essence has been decanted into a test tube, becoming part of a vast experiment whose outcome remains unclear for the time being. We can cure diseases and help people avoid suffering, but we also risk accidentally creating new problems. Describing everything that's happening in the field of gene technology – all the remarkable, wonderful, challenging research, and all the debates about where the results may lead – is rather like being a grizzly bear standing in the middle of a waterfall trying to catch salmon. There are so many exciting flashes of silver that you can't keep track of them all; it's easy to be overwhelmed by the torrent of new discoveries churning around you. Picking out a few, however, can help us get an idea of the big picture.

Everything suggests that gene technology is set to enter our lives in amazing and revolutionary ways. It will save lives and reduce suffering. At the same time, it will pose existential and ethical questions, forcing us to reflect on just who we are, and where we draw the boundaries of what is human. But in the midst of all these uplifting and transformative developments, it will also seep into the most humdrum corners of our existence, into areas you hardly even notice or think about. Isn't that a thought-provoking idea?

P.S. If you are as keen as I am to follow up details and examine things in more depth, why not take a look at the notes, sources and tips for further reading at the end of this book? The endnotes provide background information that goes beyond the scope of this book, as well as plenty of references to interesting reading, which naturally includes my sources.

1

A Modern Prometheus

Making his way to the podium at a high-level conference in Hong Kong in November 2018, He Jiankui is visibly nervous. An indignant murmur rises from the audience. A few seconds ago, the moderator called on listeners to 'let him speak without interruption' and to 'remember we are here to listen to what he has to say' – a very unusual request when a scientist is about to present his results.

The moderator, a prominent geneticist, also emphasizes that when the organizers booked He Jiankui many months previously, they had no idea what results he was going to present. As for He Jiankui himself, he is clearly proud of his achievement.[1]

'If this were my child – if I were in the same situation – I would give this a try,' he says later, during the Q&A session.

* * *

In 2018, 200 years after the publication of *Frankenstein*, Trinity College Dublin celebrated Halloween with a reading of the whole novel. At more or less the same time as Mary Shelley's words were ringing out in the venerable university, twins 'Lulu' and 'Nana' – the first ever gene-edited babies – came into the world.[2] This was not the first time scientists had intervened in the genetic material of unborn babies, but never before had anyone deliberately altered a specific gene.

He Jiankui took eggs and sperm from a couple who wanted to become parents and performed conventional IVF. But before the fertilized eggs had begun to divide, he injected them with the 'gene scissors', CRISPR-Cas9, to alter a specific gene in their genetic material.

The gene that He Jiankui chose to alter, known as CCR5, is responsible for a tiny part of the immune system. Just like other genetic material, this gene sometimes undergoes a small change or mutation, and the new, mutant variant is passed on from parents to children. The result is that people carry different variants of the same gene or genes, depending on what they happen to have inherited from their parents. In most cases, tiny mutations of this kind make little difference; the genes function just as normal. Some mutations can cause problems that result in disease. But in certain rare cases, mutations can actually bring about a small improvement.

Some people have a mutant variant of CCR5 that seems to protect them against HIV. This variant is quite common in Europe, where about 10 per cent of the population have inherited it from one parent. However, to be protected against HIV you need to have inherited it from both your mother and your father, and that applies to only a very small percentage of people.[3]

Given that HIV has only been infecting people since the twentieth century, this mutation of CCR5 can't have become widespread just because it protects against that specific virus. It's been speculated that it may also have provided protection against other diseases in the course of history, such as bubonic plague or smallpox. However, some studies suggest that people who have inherited this particular mutation are also at greater risk of contracting other illnesses, such as West Nile fever and flu. The international scientific community doesn't yet know quite how it affects the body, apart from reducing the risk of an HIV infection.

Though we humans have been gene-editing animals for some decades now, there has been a near-total consensus among scientists that we shouldn't interfere with human embryos that are going to be brought to term. To grasp why this is so much more revolutionary than other gene technologies, we need to understand a basic issue: there are different categories of cell.

* * *

The human body contains over 37,000 billion cells, ranging from the specialized light-sensitive receptors in our eyes to the rectal muscles that move what remains of the food we've consumed to its final destination.[4] But gene technology distinguishes just two types: somatic cells and embryonic cells.

Nearly all cells are somatic. The word comes from the Greek *soma*, meaning body. Somatic cells belong to us alone, as individuals: our children don't inherit cells from our nose or heart. That means that the risks associated with applying gene technology to cure a tumour, correct a failing eye or support a damaged liver always come down to the danger of harming the individual patient. While such risks may be considerable, they are always weighed up against the pain and the difficulties the patient suffers as a result of the disease which the medical practitioner is trying to cure. They are the same risks that proved so devastating for Jesse Gelsinger – tragic in their effect, but confined to a single person.

However, there are major existential issues at stake if, like He Jiankui, you alter embryonic cells: eggs, sperm and the very first cells that form once an egg has accepted a sperm and begun to divide. At this stage, interventions that modify genetic material can have a huge impact. They may cure terrible genetic diseases, or, possibly, reduce the future child's risk of succumbing to everything from Alzheimer's to heart attacks. However, editing embryonic cells has two major consequences. First, the genetic modification will be present in every cell of the body into which the embryo develops. The change effected by He Jiankui will be with

'Lulu' and 'Nana' all their lives, from birth to adulthood to menopause, and in old age.

Second, the edited genes will be passed on to the next generation. A gene-edited child who becomes a grandmother can pass on the altered gene to her grandchildren and their grandchildren's children. There is a chance – and a risk – of changing the whole future of humankind. That was why the audience murmured and the cameras flashed when He Jiankui announced the girls' birth.

* * *

He Jiankui had contacted an organization supporting people living with HIV in China, BaiHuaLin, to request help with finding couples consisting of a would-be father with HIV and an uninfected would-be mother. The aim was to protect any future offspring from being accidentally infected by their father in daily life, and from the discrimination and stigma suffered by many people with HIV in China. So he was looking specifically for couples in which the would-be father felt vulnerable and discriminated against because of his infection. To protect an embryo against HIV infection from a father with the virus, the sperm of the father-to-be are 'washed' before insemination to rid them of the virus. He Jiankui also followed this procedure. Three couples decided to take part in the experiment, though one pair later withdrew.[5]

The parents-to-be had a choice between modified embryos, in which He Jiankui had attempted to edit the CCR5 gene, and unmodified ones. Both couples opted for gene-edited embryos. By the time He Jiankui spoke at the conference, the second woman was pregnant with a gene-edited baby, who was presumably born in 2019. The reason the child's fate remains unknown will soon become clear.

When He Jiankui announced his results, it turned out that he hadn't succeeded in creating the precise variant of CCR5 that scientists know protects against HIV. Rather, he had produced new mutations which may – or may not – have the same effect. In the case of one of the twins nearly everything went as planned, so that the same mutation is to be found in all the child's cells, at least in theory. With the other, however, something happened that's common when humans try to gene-edit animals: some cells were changed, but not all of them. Presumably that was because the egg had already started to divide before He Jiankui edited the gene. As a result, the child's body is now a mosaic of modified and unmodified cells. It isn't clear whether this will protect her against HIV, or whether it will have any other effects in the course of her life.[6] He Jiankui has been criticized for failing to halt the experiment when he saw that the wrong mutation had taken place and that not all the cells had been altered.

There's also a risk that the CRISPR 'scissors' may have cut genetic material in several places, modifying some other gene in the girls' bodies. It's incredibly hard to identify tiny changes in genetic material. He Jiankui claims to have searched for unintended modifications without finding any, but it's impossible to be sure.

There are no international laws to prevent scientists (or countries) from gene-editing embryos. The 1997 Oviedo Convention, signed by 30 countries, bans inheritable genetic modifications, but countries such as the UK haven't signed it, claiming that it is too restrictive.[7] Conversely, Germany has refrained from signing because it regards the convention as too permissive.

So views on gene-editing vary, and the stance of each individual country is a matter of national law. However, in 2015 some of the world's leading geneticists took the initiative of reaching a kind of gentleman's agreement. The gist of it was that they would edit embryonic genes only in the context of basic research to deepen understanding of diseases and foetal development. Thus only embryos that would never develop into babies could be modified. The scientists concerned decided that experimental gene-editing of foetuses that would be brought to term was irresponsible until the safety issues had been resolved, and until there was broad societal acceptance of such experiments, and the scientific community could participate in an open process. These were idealistic but reasonable aims. For the time being, the watchword was 'No tweaking!'[8]

If the truth be told, the conference in late 2015, when the guidelines were drawn up, was a hastily arranged affair. The new gene technology was still a novelty. Although it had always been clear that the method could be used on human embryos, most of the scientists assessing the situation thought that stage was a fairly long way off. This was partly because of existing legislation, but largely because science sometimes develops at a very uneven pace.

But this time things moved ultra-fast. Immediately after the publication in 2012 of the first studies showing the workings of CRISPR, the 'gene scissors', scientists in China began to try gene-editing human embryos. The first scientific article to show that this was possible appeared as early as spring 2015.[9] The study in question was the work of Chinese scientists, who had tried to edit a gene responsible for thalassaemia, a hereditary blood disorder common in South East Asia and the Mediterranean region. This experiment was sufficiently impressive, and aroused sufficient concern, to push the scientific conference mentioned above into drawing up guidelines the very same year. Since then, quite a few scientific papers on modified human embryos have been published. After the first few that came out between 2015 and 2018, they have been appearing with increasing frequency. Soon after the first results emerged, scientists in countries including the United States, the UK and Sweden began applying CRISPR to embryos to clarify certain scientific issues. However, most results have come from China, where the technique has been further developed and a number of problems resolved – though application has been restricted to basic research, without any children being born as a result. Up till now.

* * *

Beijing's Forbidden City lies in a hutong district, an extensive area of one-storey dwellings built around small inner courtyards, forming an intricate system of alleys and

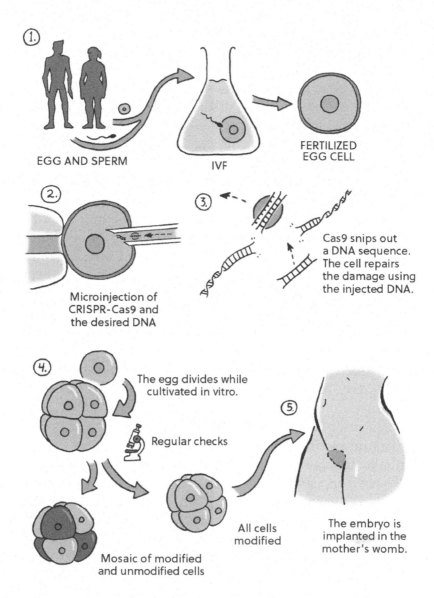

1. EGG AND SPERM

IVF

FERTILIZED
EGG CELL

2. Microinjection of
CRISPR-Cas9 and
the desired DNA

3. Cas9 snips out
a DNA sequence.
The cell repairs
the damage using
the injected DNA.

4. The egg divides while
cultivated in vitro.

Regular checks

Mosaic of modified
and unmodified cells

All cells
modified

5. The embryo is
implanted in the
mother's womb.

Any genetic modification of the embryo must take place before the
fertilized egg starts to divide. Scientists inject CRISPR-Cas9 into
the cell, where the CRISPR system can cut out or edit chunks of
genetic material. By the time the egg starts to divide, all its cells
will have been modified. However, occasionally something goes
wrong, resulting in an embryo that is a patchwork of modified and
unmodified cells.

narrow streets. Parts of this quarter date back to the fourteenth century. Despite the scale of rebuilding and social change, some of the most central hutongs remain as they were, and they teem with life. People sit at tables playing games, others lounge in courtyard doorways, smoking or keeping an eye on children at play. Strolling through the alleys gives one a sense of timelessness in a city that's 3,000 years old. I am here to investigate how far Chinese scientists have advanced with the very latest developments in gene technology.

If you walk eastwards from the Forbidden City and stroll through the picturesque hutongs for a good half-hour, you'll reach a newly built shopping centre, cheek by jowl with the old quarters of town. It looks as if a highly sophisticated spaceship has landed in the city, a vessel built by wealthy extraterrestrials with a finely developed sense of the aesthetic. Huge oval beehives of white concrete and dark glass are linked together by gently curving footbridges, illuminated fountains and sculptures. These edifices, teeming with stores and boutiques, are a magnet for families on shopping trips. This centre provides a glimpse of the new China, a country that makes a European wonder whether she's accidentally time-travelled into the future. In a Starbucks cafe inside the shopping centre, I meet Tang Lichun, one of the first scientists to succeed in modifying a human embryo in a laboratory.

'We began this project as early as 2014, very early. And in the beginning we had many difficulties; it didn't work on the embryos. [...] We tried and we tried [...] and got results. It was very exciting,' says a beaming Tang Lichun when I ask him to tell me about the experiment.[10]

The technical problems he and his colleagues encountered meant they weren't the very first to demonstrate that the new technology could be used to gene-edit human embryos. But they were the first to show that it can also be applied to healthy embryos – embryos that could develop into babies – although Tang Lichun only allowed the tiny cell clusters to develop for a few days.

When reporting on new, groundbreaking research, the people I meet are nearly always professors, the scientific elite at the peak of the hierarchy. It's rare for me to meet the people who stand at lab benches, among test tubes and Petri dishes. That makes it a particular pleasure to meet Tang Lichun, whose experiments on embryos were part of his doctoral research. When I ask him which aspects of the study he was involved in, he replies, 'All of them.' Although the study was successful, he thinks it will be a long time until embryos can be reliably gene-edited to prevent disease.

'Maybe the research can be used clinically in future, but not so far,' he says, explaining that there are many technical problems with the technology, as well as unknown factors. Rather, he thinks gene technology has potential as a useful tool for research into embryos. There's so much more to discover.

I ask him how he thinks the general public should view CRISPR, the gene scissors. In contrast to the rather dismissive replies I've received from various professors on my journey, he reflects at length before finally replying: 'This technology indeed gives some people hopes that diseases that were untreated in the past can be cured. [...] This gives them hope, but they should also know that the wait is long.'

He thinks that until CRISPR technology improves sufficiently for users to be able to avoid undesirable side effects, it won't be feasible to use it for medical purposes.

* * *

This caveat, that the technology needs to improve greatly before it will be safe, goes without saying. Most of the world's scientists agree we haven't yet reached that stage. Yet He Jiankui defended his decision from the rostrum.

'If you have any friends or family members with a disease or disorder, you'll understand. They need help. And we now have a technology that can help millions of families with hereditary or infectious diseases,' He Jiankui announced. He hopes that 'Lulu' and 'Nana' will soon be followed by more gene-edited children.

'I'm proud, incredibly proud, of this particular case,' He Jiankui continues. 'The father had lost the will to live, but now the couple have healthy children, he feels able to take care of his family. I genuinely believe that what we have here is essential for the millions of children born to parents with HIV, since we don't have any vaccine against it.' Yet many scientists question whether this was the right disease to start with, whether it's really so serious as to constitute an acute medical need. After all, there are other ways to protect children against their parents' HIV.

After presenting his results at the Hong Kong conference, He Jiankui stayed for nearly an hour to take questions. The scientists present responded in essentially the same way as the rest of the scientific world. They were virtually unanimous in their condemnation.

Not only had He Jiankui crossed a scientific line, he'd done so in secret. He hadn't told his university about his experiments, nor had he applied for authorization to conduct his studies, at least not according to the Chinese authorities. He took a sabbatical in February 2018 to carry out the necessary experiments himself, using his own company as a base. Moreover, the results weren't published in any scientific journals or peer-reviewed. The Chinese authorities have now brought down an iron curtain around the experiment and all those involved. When the result came to light, China's vice minister for science and technology, Xu Nanping, said the experiment had 'crossed the line of morality and ethics [...] and was shocking and unacceptable'.

He Jiankui disappeared after the conference and hasn't been seen since. But on New Year's Eve 2019, it was reported on Chinese state news that he had been sentenced to three years' imprisonment and fined 3 million yuan. Two of his colleagues received shorter sentences. Investigations are also ongoing in the United States to find out whether any American scientists were involved.[11]

Despite the sentence, most of the circumstances surrounding the experiment remain unclear, and although several years have passed since the children's birth, there hasn't been any further information about the fate of the twin sisters or the third gene-edited baby, who was presumably born a couple of years ago. A number of investigative journalists are trying to find out what exactly happened, how the studies were financed, and what the scientific world can learn from these cases. China's unwillingness to communicate with journalists and scientists has made matters still more difficult.

Yet these experiments have already inspired a number of people to follow suit. Russian molecular biologist Denis Rebrikov, who works for Russia's largest fertility clinic, has applied for authorization to conduct similar experiments. He wants to edit the same gene as He Jiankui, even though the latter's experiment wasn't completely successful. He also wants to try to edit another gene, one that causes deafness.[12] Although Rebrikov seems to be the only person to have stated openly that he aims to follow in his Chinese confrère's footsteps, there are sure to be more very soon.

There are likely to be more gene-edited babies on the way by the time you read this, whether the process is public or secret.

He Jiankui seems to have rushed the experiment through so as to be the first. Already, it would be possible to do it more safely. Just like virtually all other scientists, He Jiankui claims he's against using gene-editing to 'enhance' human beings. The technique should only be used for 'curative' purposes, he says. But the question is exactly where to draw the line. You could say the twin sisters were 'enhanced', as the aim was to give them protection rather than to cure them.

'We wanted to begin with a straightforward disease that's been studied in depth, so that in future we can continue with more complex disorders involving more than one genetic modification,' repeats He Jiankui when he's asked, again and again, why they picked that particular disease, those particular would-be parents, this particular time.

Even He Jiankui's critics include a number of scientists who see his experiment as 'a misstep on the right path', and who hope to see more gene-edited children soon. Several scientists have also stressed the need to develop just this kind of technology for certain types of disease where there is no alternative. They say this research must be allowed to continue, regardless of whether the pioneers acted correctly or not. Another speaker at the conference, geneticist George Q. Daley, said in his address: 'As a species, we need this flexibility in order to withstand future threats; we need to take control of our own heredity.' Others have spoken of 'a technology that could save our species'.[13]

CRISPR technology has given us a brand-new set of tools to modify our genes, but how good is our gene instruction manual? Despite the risks and uncertainties, there are parents who dream about altering their future children and clinicians who hope to make money out of the new technology.

2

Build Your Own Baby

In South Korea, having a baby is clearly a big deal. The Seoul Metro reserves special seats for pregnant women. Upholstered in mauve and pink, each has a round, blue cuddly toy the size of a football attached to it. Why an expectant mother needs a cuddly toy is something of a mystery, but apparently that's how it is. Above each seat there's a large picture of a smiling, blue-haired woman with a heart in her midriff. The public address system announces that these seats are only for pregnant women, including those whose pregnancy isn't yet advanced enough to be visible. Quite a few women on the metro sport a large round badge featuring the same blue-haired woman in miniature. You can get one by showing the authorities a medical document certifying that you are pregnant and therefore entitled to a special seat.[1] I never see a passenger without a badge in any of these seats, not even when the carriage is chock-full.

It's hardly surprising that South Korea sets such store by mothers-to-be: the baby shortfall is starting to pose problems for the country. My week in Seoul coincides with the announcement that South Korea's birth rate has now fallen below one child per woman for the first time. The downward trend has continued ever since.[2] One of the reasons for the baby deficit is that South Korea remains a very patriarchal society. Many women postpone marriage and childbirth to avoid discrimination on the job market. Although the country has made efforts to adapt its policies so as to encourage more women to have children, it isn't clear how effective they have been.[3] The average age of a first-time mother is now just over 31.

Economists predict huge problems in future as the population needed to support those of pensionable age declines. There are other risks, too: empty classrooms and shrinking armed forces. Japan and many other countries face the same issues. Half the world's people now live in countries where the average number of children per family is below the rate needed to maintain the population at the existing level. Where population increase continues, it can be attributed mainly to greater longevity and the fact that girls already born are growing up and having children.[4]

There are many reasons why women worldwide are choosing to have children later, or not to have any at all – at least, those women who are able to decide for themselves when, and how often, they want to give birth. Most of those reasons are positive: women are pursuing education and careers. More women can now provide for themselves, so they can choose whether or not to have a partner – and if so, when – instead of feeling under financial pressure to do so. So everything suggests that the reason for these figures is that people are making rational decisions, in Seoul just as elsewhere in the world. However, when women choose to have children at a later age, more of them face a struggle to conceive.

Maybe that's why Seoul has now become one of the top locations for fertility treatments of various kinds. And it's not just Koreans who are availing themselves of assisted reproductive technology: every year, thousands of couples come to Seoul for help with creating a family.[5]

With a little time to spare in Seoul, I decided to visit one of the now world-famous clinics, the CHA Fertility Center next to one of the capital's main railway stations. CHA Health

Systems, to which it belongs, is a company with four clinics in South Korea and one in Los Angeles. The Fertility Center resembles a designer hotel more than a medical facility. The outer section features attractive sculptures, little fountains and sofas in pink.[6] Through the glass walls in the lobby, you can glimpse large metal tanks for freezing embryos. Tastefully lit and positioned against a mauve background, they are as far removed as possible from a laboratory aesthetic. The ambience of the clinic's inner sanctum is cosy and comforting, with discreet music, designer sofas in pastel shades and undulating walls in wood and metal.

The sales representative who shows me around tells me the design is intended to make women feel well cared for and to give them a sense of privacy. There are no open spaces where one might feel exposed, or even any sharp corners that might be perceived as stress-inducing. The capacious armchairs have large wings, allowing the occupant to drift away into her own private world. 'It's designed to be rather like a hive,' says my guide. There is also a discreet urology department, where male patients with possible sperm-related problems are examined. This has a separate entrance, enabling them to avoid embarrassment and remain isolated from the female patients.

There are plenty of people here, about 20 women and a few couples. Ten or so clients come and go while I'm waiting. The meandering layout of the corridors makes it hard to form an impression of the clinic as a whole, which is the idea. You're supposed to feel cocooned.

More and more women are choosing to spend a month or more here to be helped to conceive. A course of treatment costs somewhere between 6,000 and 10,000 US dollars, and all the doctors at the centre offer online consultations. I'm handed a thick brochure about the centre's history and the kinds of treatment it offers. The sales rep tells me they have numerous foreign clients and can cater to their needs in several languages. As well as providing medical care, they can help with hotel bookings and airport transfers. She shows me a list of their doctors. They all speak English, she says, but some are more fluent and 'culturally' attuned to foreign clients, as she puts it, than others.

Now that He Jiankui has taken the first step with gene-edited babies, it's probably places like this one – luxurious, cosseted, discreet – that will offer the first commercial gene-editing treatments. That may not happen in this particular facility, as South Korea has fairly strict regulations on IVF and related treatments, but it's likely to involve a clinic with more than a passing resemblance: perhaps on a tropical island, or in a big shopping mall in some country where the laws are more relaxed. Mexico, perhaps, where the rules on various types of medical treatments are laxer, making it attractive both to charlatans and to serious researchers keen to try out new methods. So far, Mexico hasn't brought in any explicit ban on genetic modification.[7]

* * *

A few days before my visit to Seoul, I visit Sapporo in northern Japan to meet Ishii Tetsuya, a bioethicist at Hokkaido University. He takes me to a university cafeteria, where we have lasagne for lunch – which seems rather a pity in Japan, of all places. But the menu offers only Western dishes, and I assume that good restaurants are as rare here as they are on most campuses. Or perhaps this is just a courteous gesture on my host's part.

Ishii Tetsuya is a leading world expert on how gene-editing in human beings will affect us and our society. He focuses above all on the foetus and gestation, on what happens when we start tinkering with the DNA of unborn children.[8] The main issue here is who exactly is to take such decisions. So I ask whether he thinks we're going to start gene-editing our children. Will this go beyond individual scientific experiments?

'Yes, that's unavoidable,' he says. 'We lack enough wisdom, but we try to use new technology to become the parents of a child with a trait we want: health, maybe, or a visible trait like hair or eye colour. I think that kind of modification may also be possible in future.'

During our conversation, he comes back to the same point several times – the issue of consent as it applies to children born with genetic modifications. The point is that the children themselves are not in a position to consent to any such change.

'It is important to remember that you and I were born without our consent. It is our parents who determined our fate. With gene-editing in human embryos, it will be the parents and doctors that give their consent, not the children. And this is a very big, important thing – a predicament for the children born this way, and for future generations.'

Tetsuya's English is slightly hesitant and not always easy to understand. He expresses himself with typical Japanese understatement, and his criticisms and judgements are precise and carefully considered. This is clearly a subject close to his heart. He is concerned to avoid any misunderstanding or over-interpretation. You can see he is deeply invested in the subject: he gestures while speaking, sometimes pausing awhile to really think through what he wants to say, smiles and laughs. The word he repeats most is 'troublesome'. He is at pains to point out that this is a complex subject and that we live in uncertain times.

'But I think the most important thing is consent. If a tragedy happens, I think the child will accuse the physician or its parents. That's extremely troublesome,' he says. 'What about a child born with a genetic modification that goes wrong, or one they just don't want? It's easy to see how that could damage a family deeply, for a lifetime. Some pregnant women may have to terminate their pregnancies, depending on the genetic condition of the foetus. That would be a tragedy,' he continues. 'If a child is born with congenital anomalies, it is very, very tragic.'

Helping himself to some lasagne, Ishii Tetsuya continues: 'This technology will have a huge impact. It may be a good effect, but in some cases, probably, say, one or two in ten cases, there will be a devastating effect.'

This is an argument that could equally well be used in defence of gene-editing. A child born with an avoidable disease might well accuse its parents of failing to do anything to prevent it – of not opting for gene-editing.

* * *

There are two problems people hope to solve by gene-editing embryos. First, they want to avoid babies being born with serious genetic diseases. Gene-editing offers a possible way to save children with such serious genetic anomalies that they die in the womb or immediately after birth under present conditions. Second, gene technology offers hope to women who face difficulties in conceiving. Both types of genetic intervention have in fact already been applied – and this happened quite some time before He Jiankui made his announcement.

Two years earlier, in 2016, a Jordanian woman with a mitochondrial disease gave birth to a gene-edited baby in Mexico, with the assistance of an American doctor. Her pregnancy represented an attempt to overcome both a genetic defect and a fertility problem. Yet the headlines about this breakthrough were silent on the subject of genes. Instead, the newspapers devoted page after page to the news that a child with three parents had been born.

An interesting, rather intriguing fact about us humans is that our cells contain two quite distinct types of DNA. On the one hand, there are the chromosomes located in the nucleus of each cell, which are what we usually mean when we talk about our genes. Just about everything you've ever learned about DNA has to do with the cell nucleus.

The other type of DNA, on the other hand, tends to be neglected – but it's at least as important. Inside the cell, but outside its nucleus, are all the chemical and biological systems that enable our bodies to function. This is where proteins, hormones and enzymes are produced. This is where phenomena like circadian rhythms, fat storage and the sensation of hunger are regulated. In short, this is where all the action happens. While the cell nucleus contains the instructions, it is outside the nucleus, in the cytoplasm, that processes actually take place. The cytoplasm is the factory floor of the individual cell, and of the body as a whole. It contains tiny clusters called mitochondria that function as the cell's energy generators: thanks to a complex chemical process, they furnish the cell with a constant supply of fuel that covers all its needs. Our cells wouldn't survive for even a few minutes without their mitochondria. But the really fascinating thing is that mitochondria – unlike everything else in the body – have their own genetic material. They have their very own DNA, which never combines with the DNA in the cell's nucleus.[9]

No one knows quite where mitochondria come from, but the most likely theory is as follows. Once upon a time, about one and a half billion years ago, a single-celled organism happened to ingest a bacterium. At that time, the only organisms in existence were single-celled organisms, which would subsequently evolve into humans and animals, and different types of bacteria, which have continued to be amazingly successful. To be honest, the difference between our forerunners and the bacteria wasn't that big. Our single-celled ancestors had a cell nucleus, and they could ingest bacteria. So the fact that one of them gobbled up a bacterium wasn't a particularly exciting event. In this case, however, the bacterium survived inside the single-celled organism.[10]

The bacterium absorbed nutrients from the fluid inside the cell and excreted by-products which the cell could use. It also benefited from protection against being devoured by some other organism. It divided and multiplied within the cell, and when the single-celled organism grew and divided, both parts had copies of the bacterium. After that the details become slightly hazy, but for some reason those of our single-celled forerunners that contained these bacteria did better than the others, and they quickly proliferated. Maybe the bacteria helped them break down some nutrient which they couldn't digest on their own. The bacteria had started out as parasites and free-riders, but in time they became essential to the cell's continued existence. They were so comfortable inside it that they soon lost the ability to survive unaided. After thousands of generations, bacteria and cells became so dependent on one another that they were inseparable. These bacteria developed into our mitochondria, while the single-celled organism prospered so well thanks to its new partner that it took over the world. All the animals, plants and fungi alive today have mitochondria. Although they are an essential part of us, they retain a small amount of the genetic material from the original bacteria, and they can divide autonomously within our cells. The mitochondria lie outside the cell's nucleus and we inherit them only from our mothers; when an egg forms, it incorporates the mother's mitochondria.

* * *

This would be a scientific aside if it weren't for one thing. Mitochondrial DNA, just like the DNA in the cell nucleus, can mutate. These mutations can result in serious disorders, and can sometimes even be fatal. About one child in every 5,000 is born with a mitochondrial disorder, and for many of them there is no effective remedy. These diseases can vary from the mild to the life-threatening.[11]

When a woman has trouble conceiving, one possible reason may be that the mitochondria in her eggs have been damaged in some way. This has been known for quite some time. Doctors in the United States began experimenting with a new way to help women with infertility problems back in the 1990s. They transferred part of the cytoplasm of donor eggs to the eggs of involuntarily childless women – rather like transferring part of the white of a hen's egg to another egg. This meant that mitochondria and other components of the donor egg were also transferred. It wasn't an exact science, but it seemed to work.[12]

In practice, this is a kind of genetic modification, as the medical intervention affects the genetic material of the future children. If a child receives a few mitochondria from a donor egg, it will have a small amount of DNA from someone other than its parents. Looking at things this way, the world's oldest gene-edited human is already some way into adulthood. The experiment was closed down not long after it had begun. When it emerged that some of the babies concerned bore DNA from three individuals, the US Food and Drug Administration (FDA) insisted on more documentation and a rigorous authorization system. The doctors concerned felt it was getting to be too much for them. About ten babies were born before the experiment was halted.[13]

The news spread, and doctors in India and Israel tried the technique. Fifty or so children have now been born using this method. At the same time, however, action got under way to ban it, including in the United States. There was even talk of 'human cloning', which has nothing to do with the scientific facts.

Some doctors continued to experiment, however, and in 2016 the first baby was born through a technique involving the transfer of nuclear DNA, together with the rest of the nucleus of the egg, from a prospective mother to a donor egg with healthy mitochondria from which the nucleus and its DNA had been removed. This is a more complex process than transferring a little fluid from one egg to another. The next step is to fertilize the egg with sperm from the prospective father, after which it is implanted in the uterus of the mother-to-be. This is a far clearer example of genetic modification than the earlier experiments, as all the mitochondria come from the donor egg.

The mother who bore the first baby created this way had Leigh syndrome, a serious mitochondrial disorder affecting the nervous system. The couple had already suffered four miscarriages and lost two young children, but after the treatment things seem to have gone well. This new technique is legal in Britain, and Australia has recently announced that authorization is being considered. Other countries are likely to follow suit before long.

However, for the time being this method remains either banned or unregulated in most of the world. Sweden, for instance, does not allow mitochondrial treatments, at least not yet. But those would-be parents who can afford it can travel to clinics willing to provide treatment – in Spain, Russia, Albania, Lebanon or Israel, for instance. Doctors at a Ukrainian clinic offering this technique have said in interviews that they have helped a dozen or so women to conceive, including one from Sweden.[14]

Bioethicist Ishii Tetsuya sees strong similarities between mitochondrial treatments and processes that involve modifying the DNA in the cell nucleus: partly because these techniques can help would-be parents who cannot have healthy children by any other means, and partly because they provide a way to reduce the incidence of diseases that cause terrible suffering.

'If a couple are at great risk of transmitting pathogenic mutations to their children, it might be advisable. But we should also bear in mind that donor eggs or sperm offer a possible solution, so it all depends on what value people attach to their own genes,' he says. Ishii Tetsuya is at pains to stress that he knows about the problems associated with donor eggs and sperm, but that they nonetheless offer a better solution.

There are alternatives to gene-editing for most genetic diseases. Many countries, including Sweden, provide genetic advisory services for couples who are thinking about having children despite a genetic disease in the family. Where there is a risk of a serious disorder, the most usual procedure is IVF treatment resulting in several embryos. The doctors then conduct genetic tests on the embryos to find out which ones carry the genes causing the disease, and select those that haven't inherited the defective genes. Another alternative to gene-editing that is set to become more common is gene therapy, which involves treating children after birth.

Despite the risks, and even in the event of tragic foetal damage and miscarriages, Ishii Tetsuya foresees a shift towards deploying the new type of gene-editing used by He Jiankui:

'If a new technology can correct mutations, I think countries like the UK will go a step further. They will consider authorizing that technology to prevent serious disease, again with the emphasis on *serious.*'

IVF treatment was initially thought unnatural and immoral, and families were encouraged to adopt instead. Yet today it has resulted in the birth of over 8 million children, and though tough and gruelling – both mentally and physically – it is now quite uncontroversial under most circumstances. It's a technology that has helped many families to have healthy, wanted and loved children.

Ishii Tetsuya assumes that the gene-editing of embryos will eventually be regarded in the same way as IVF treatment, donor eggs and sperm, and the option of selecting a baby's sex, which already exists for those willing to pay. He believes there will be a market among wealthy prospective parents who have already decided to use IVF and want to protect their children from a genetic disorder or give them an added genetic advantage.

Reproductive tourism is already a reality, and the industry looks set to expand further in future.[15] People travel abroad to obtain IVF treatment more cheaply than at home, but also to get around restrictive laws. Some countries won't allow couples, or single women, to use donor eggs or sperm. Some prohibit fertility services for same-sex couples. Others have regulated surrogate motherhood by restricting the permissible sources of eggs and sperm.

It's only just become legal in Sweden to carry out IVF treatment using both a donor egg and donor sperm. Before that, either the sperm or the egg had to come from one of the parents-to-be. The existence of well-heeled couples who want children, coupled with a legal situation that varies from one country to another, has created a market.

There's no doubt that the first services on offer will involve editing out genetic diseases. Clinics will offer to remove the defective genes behind disorders like Huntington's disease, haemophilia or thalassaemia. In the wake of He Jiankui's breakthrough, it has emerged that he had plans to set up a 'genetic medical tourism' business. This would have offered treatment in Thailand or China, and it would have targeted an 'elite clientele'.[16]

'Reproductive tourism is a big problem,' says Ishii Tetsuya. 'So now the situation is that new technologies are appearing that involve modifying cells, and some countries are responding, but others aren't. Many couples go abroad for fertility treatment, so the situation is very complex.'

Many of the laws concerned (such as the fact that it's illegal in Sweden to choose the sex of your baby) are reasonable, but the regulations have also created a demand for donor eggs, for example, without there being careful checks on where they come from.[17] Another problem is the checks on whether clinics really provide the treatments they claim, and how careful they are to ensure that the methods they use have been rigorously tested and fully developed before they start offering them to paying clients.

After tackling the most obvious genetic disorders, the next reasonable step is to reduce the risk that the future child will suffer from other diseases. Clinics could follow in He Jiankui's footsteps by trying to minimize a child's risk of being infected with HIV, or by editing gene variants that heighten the risk of diseases like breast cancer or Alzheimer's. Yet science lags some way behind people's hopes. The debate that ensued after He Jiankui's experiment showed that we lack sufficient knowledge about all our genes to make such changes safely – for the time being.[18]

Another possibility – still at a very early experimental stage – is helping same-sex couples to have children. There are many studies under way into whether stem cells can be converted into eggs or sperm, which would make it possible for two women, or two men, to become genetic parents. For various reasons, it seems to be harder to produce eggs than sperm, so a male couple would still need donor eggs. But 2018 saw the birth of baby mice with two mothers.[19] This experiment, like many other genetic breakthroughs, took place in China. The young mice were healthy and able to have healthy offspring of their own. Similar experiments involving two male mice have proven unsuccessful so far – and there's a big difference between mice and men.

However, developments could also take a similar course to the United States, where the breakthrough in genetic technology has led to more stringent legislation. Some observers think many countries will change their laws to make gene-editing in humans illegal now that the first gene-edited babies have been born – perhaps so soon that there won't be enough time to consider the possible advantages. What would happen then, as with mitochondrial treatments, is that the technique would move from big laboratories in big countries to smaller clinics in countries with laxer laws – and to patients able to pay for the privilege.

'The next step after tackling disease is hard to predict. Maybe "designer babies". But there's also a risk that that may lead to problems like congenital anomalies or miscarriages,' says Ishii Tetsuya.

After curing disorders or reducing the risk of their occurrence, clinics with ambitions and clients with the wealth to indulge their own preferences will probably head out into deeper waters. The idea that there are clear links between particular genes and particular traits can still be seen in the way news about genetics is presented: claims are made that scientists have found a gene for addiction, a gene for obesity, a gene that makes us thin, a gene that makes us more inclined to take risks, and so on, even though the truth is far more complicated. One can easily picture enabling would-be parents to tick boxes for musical talent, creativity or a penchant for long-distance running.

Scientists and politicians are concerned that 'designer babies' will be born in future, with parents cherry-picking their future children's traits from an attractive menu – and not just to avoid disease. After all, it's already possible to make that sort of choice thanks to donor eggs and sperm. (Shall we use sperm from a tall man, or a short one? Eggs from a highly educated woman? Blue eyes, brown eyes, how about green?) So what happens when the option of editing the baby's genes as well enters the picture?

Discussions about designer babies nearly always end up focusing on intelligence. Can we use gene technology to create hyper-bright children? What would happen if some mad scientist were to start creating a race of superhumans? Could we become clever enough to find solutions to all our problems? Would that be a good thing – or a disaster? The notion that we might somehow be able to enter the human body and flip the switch labelled 'intelligence' is one of the greatest temptations of gene technology, but it's also the aspect that scares people most.

Intelligence is complex. First, scientists aren't sure exactly what the term signifies, or what an intelligence test actually measures. Second, it's clear that intelligence is partly hereditary: individuals with high IQ scores have children who also score highly. This is true even in cases where the children grow up in a family other than their biological one. But intelligence is also affected by various aspects of a child's upbringing, ranging from nutrition and vitamin intake to the number of books in the home and access to education. Just what proportion of intelligence is hereditary and what proportion is determined by a person's environment remains a much-discussed issue among scientists. Currently, the consensus seems to be that between 50 and 80 per cent of an individual's intelligence depends on their genetic makeup, the rest being determined by other factors.[20]

Scientists have identified many genes that have some impact on a person's intelligence: a recent study counted over 500. Throughout our evolution, we humans have been subject to immense evolutionary pressures to become more intelligent. We are not the fastest predators, or the strongest, but we can hunt together in groups. And living in groups places particularly high demands on intelligence. We have evolved to be constantly aware of each other and our mutual relationships; we have learned to understand and to cope with all the affinities and the conflicts that come and go within groups of people. There have always been advantages in telling stories or in remembering where sweet-tasting berries, or some other plant useful to the group, can be found. Our genetic material has been refined over millennia, maximizing our intelligence. At the same time, we have developed empathy and the ability to cooperate with each other: intelligence is not the sole quality that has enabled us to survive.

There have also been opposing forces at play, of course, other evolutionary forces that have resisted and pulled in the opposite direction. Babies whose brains were too big would tear their mother's birth canal, or even remain unborn. If a mother or baby dies during childbirth, the mother's genes are unlikely to be passed on. The human brain is a heavy organ requiring a lot of energy: humans need strong backs and necks and enough food. Tiny adjustments and modifications throughout evolution, together with a generous portion of chance, have created a balance. That means there is no simple, obvious gene that scientists can tinker with to boost our intellect. Tweaking one part of the genetic fabric may cause problems elsewhere, possibly in a quite unexpected area. Nonetheless, a few scientists believe it will someday be possible to select the potentially most intelligent embryos resulting from IVF or gene-editing. And there are companies that offer tests purporting to identify embryos at a particular risk of cognitive impairment, though the level of uncertainty is very high.[21]

There are a few other characteristics which may be attractive to parents-to-be in the future, traits with a more obvious genetic link. Creativity is one example. A number of studies indicate that there are gene variants that can have a real impact on creativity. That means it can make a difference to tweak a gene or two. But the downside is that there's also a clear link between creativity and a range of mental illnesses such as bipolar disorder and schizophrenia.[22] So should a parent hope to have a creative child who runs a greater risk of an extremely challenging psychological condition, or should they decide to do without the creative trait to protect their child from potential suffering?

The same sort of argument applies to many other traits, regardless of whether they are associated with just a couple of genes or with a whole range of them. There are advantages in being tall, but a tall person also runs more risk of suffering from back problems. Being genetically predisposed to impulsive behaviour can both cause problems and open up opportunities. Dark skin reduces the risk of developing skin cancer, but increases the likelihood of suffering from a vitamin D deficiency, particularly for those living at higher latitudes. And so on. The various probabilities interact in complex ways: a trait that may look good on paper can make life more difficult for those growing up in the wrong environment, and vice versa.[23] Moreover, since changes like these can be passed on, nobody knows what impact they may have on future generations.

'I'd like to ask people at any rate not to go to such clinics without thinking things through carefully. There are going to be very many troublesome situations in the future,' says Ishii Tetsuya. This development clearly worries him, so I ask what he would do if he could take a decision to be applied worldwide.

'I would introduce temporary or time-limited prohibitions, to give a chance to talk about the acceptability and the ethics of the new technologies in society. I think that would be the best solution.' He continues with an argument about why he actually believes that a permanent ban would be best, but would be hard to implement.

Taking out his laptop, Ishii Tetsuya shows me the results of various studies he and other researchers have conducted.[24] Respondents were asked if they could accept gene-editing in embryos, or in the cells of adults, as a way to prevent disease. Well over half thought both methods acceptable, although gene-editing in embryos could have far greater consequences than modifying cells.

'People seem to think it's the same thing. I don't think they're aware of the pitfalls. And I think it's the same in many countries. We need much more dialogue. There are big misperceptions about the facts, the scenarios and the purpose,' he says. 'But again, we really have to think about the aspect of consent. We can consent, but the child can't. Maybe that's the biggest problem of all.'

Like many scientists, Ishii Tetsuya is concerned more than anything by the fact that people don't seem to be giving much thought to these issues, about the risks and consequences of developing this technology. Instead, they just get used to it and quickly adjust to the new normal.

'This is common in modern society,' he observes. 'We enjoy technologies that may potentially have a harmful effect. But we need to maintain a good distance from new technology. In some cases I think we need less distance, in others there should be more. But it is not easy. We have to change our thinking. I also respect the effectiveness of gene-editing, which can be accurate and very efficient. But it has an off-target effect.'

Looking at the broader perspective, he also sees a range of societal risks. Now that there are so many fertility clinics in Japan, fewer people are adopting, and more children – especially those with congenital anomalies – are growing up in children's homes. This has gone so far that the Japanese government has started running information campaigns and taking other action to encourage more couples to adopt. Moreover, treatments like this, costly and complicated as they are, are liable to amplify class distinctions. It's already clear that only a tiny group of rich people can afford to travel abroad for treatment. Ishii Tetsuya draws a comparison with surrogate motherhood.

'Even if this proves to be a truly effective way to prevent disease, it's always going to be very expensive. I don't think it'll be possible for national health systems to cover it,' he says. 'We have no effective solution to future potential problems. I think international regulation is difficult to establish and will not be effective everywhere in the world. We have to inform prospective parents, younger people, ask them to think carefully about this.'

Walking around the clinic in South Korea, I mull over all the different images conjured up by the authors who have written about genetically engineered babies. Aldous Huxley's famous novel *Brave New World*, first published in 1932, describes long rows of impersonal incubators, pumping out de-individualized citizens with characteristics tailored to the needs of the State. That image, together with the racial biology movement of the twentieth century, has shaped our views on genetically modified humans. But it now looks more likely that gene technology will be applied in settings like these: in aesthetically pleasing, muted, air-conditioned facilities, with the help of doctors who are discreet, highly skilled and, above all, expensive. Ishii Tetsuya rounds off our conversation with a clear-sighted view of what may happen if we don't pause to consider our future prospects:

'You can control the genes of future children. You'll be like a god.'

3

Biohackers

There's a tall building in downtown Brooklyn that proves harder to locate than expected. Once I'm indoors, I think how different it all looks from the shiny start-ups with their cool logos and on-trend furnishings that are more the norm to a journalist covering gene technology. This seems to me like an older, more down-at-heel New York: the landings are littered with abandoned furniture and the lift groans and wheezes its way up the shaft, after which I have to pick my way between what looks like a plumber's fittings depot and a clutter of discarded cupboards and chairs. Finally I reach a door with a picture of the Earth on it.

Inside, there's a laboratory that's been here for less than a year. It's still a little untidy, with lots of cardboard boxes around the lunch table and the work surfaces, but the actual lab area is spotless and well organized. Four young people are decanting liquids from one test tube into another, after which the test tubes are placed inside a machine. Sitting at a desk covered in books, papers and devices of one kind and another is Ellen Jorgensen, who keeps a close eye on their work at the lab benches.

Ellen Jorgensen is the undisputed matriarch of a DIY biology movement that started back in the early 2000s. In 2010 she and a few others founded Genspace in Brooklyn. Genspace was the world's first community lab: a lab for citizen science that anyone can join. Whatever your level of training, you can get help with learning more about science and carrying out your own experiments – which can range from making your own yoghurt to gene-editing plants and bacteria.[1]

'Our question was: can you do good biological research in a more intimate way, in a more unconventional space?' Ellen says.

They were also largely motivated by the desire to improve public understanding of what it is that scientists do.

'I felt that in 2008 we were at a low point in the United States in terms of science funding and public respect for science. And I thought it was important to do something about it before the torches and pitchforks came out,' she continues.

They were pioneers, and there were no guidelines to follow, so the lab developed in a spontaneous way. People who wanted to join paid a hundred dollars a month and got access to the laboratory itself, with all its equipment, plus a range of courses and events. Informal cooperation developed as a result: all those who happened to be in the lab at a given time would help each other, which allowed them to learn about each other's projects.

Since then, many similar labs have opened worldwide. Many of their projects aim not just to teach people about science, but to try to make the world a little better. A community lab in San Francisco, for instance, is trying to find a cheaper way to produce insulin for diabetics. There are biohackers working with various types of bioprinters that can print out cells in different patterns: their aims range from producing blood vessels to creating art based on fluorescent bacteria. Ellen helped lead a study to analyse microorganisms in the Gowanus Canal that runs through Brooklyn, to assess just how badly polluted it was. Many community labs work together closely with artists, providing meeting points for the arts and the sciences, while others focus on nurturing start-ups. The answer to Ellen's question of whether it's possible to carry out innovative research in a more intimate setting is a firm *yes*.[2]

'A lot of stuff being done in these labs is at a pretty high level,' she says. 'There's a misconception that these labs are mainly amateurs who don't know what they're doing. Every one of these labs that I know of has at least one professional scientist who's there to guide the other members both from the perspective of safety and from the perspective of actually getting something done.'

What these labs aim to do is spark public interest in science in general and biology in particular. To achieve this, they hold art exhibitions and open evenings where anyone can drop in and see what's going on. The ethos of openness – as regards both access to labs and knowledge of science, as a key to understanding society – is viewed as a matter of democracy within this loose-knit movement.

'I think there's a feeling of powerlessness behind a lot of the pushback on things like GMO foods and I believe that the more we involve the public, not just in decision-making, but in the actual scientific process, the better it'll be for our society,' says Ellen.

Many labs work together with local schools: that's how the young people working at the lab benches got here. During our conversation, Ellen keeps an eye on them and interrupts herself from time to time to answer their questions and check that everything's going according to plan.

In 2017 Ellen left Genspace to set up Biotech Without Borders, which is where we are now.[3] The organization is a non-profit, most of its equipment being either donated or bought second-hand. She tells me with a chuckle that neither the heating nor the air conditioning worked when they first moved in. They kept warm coats on at the lab benches in winter, and in summer it got so hot that some of their experiments didn't work. The roof leaked, and a while back the water pipes were out of action for several days.

'Maybe some of the hacks that we had could be useful to other people,' she says. There's now a focus on opening up access to lab facilities to those under-represented in the scientific community, both in the United States and worldwide. What she really wants to do is help create community labs in Africa. She tells me she's recently attended a United Nations meeting on the contribution of biotechnology to achieving the UN's Sustainable Development Goals.

A decade has passed since the first community labs opened, and the gene technology revolution has already become central to the movement. Genetic analysis is now an everyday activity. Some labs look into which genetic variants of bacteria make the best kombucha, while others analyse the antibiotic resistance of bacteria in local rivers or streams. A striking illustration of how effective community labs can be is the German amateur sleuth who used one to identify which dog owner in his neighbourhood never cleared up after their pet – an annoyance familiar to most of us, but which few do anything about. By taking samples from dog dirt in the street and analysing the DNA it contained, he worked out which breed of dog had deposited these little treats. There being only one of that type in the area, the offending owner was swiftly unmasked.

In the last few years, since the invention of the CRISPR-Cas9 gene scissors, gene-editing has also become a standard procedure. People are trying their hand at brewing beer with fluorescent yeast or growing plants that glow in the dark. A laboratory in Paris has got bacteria to produce multicoloured ink, while one in Los Angeles is trying to alter the genetic structure of algae to produce motor fuel. Another lab is trying to engineer plants to produce sex hormones, as it's very expensive and difficult for trans people in the United States to get hormone treatment. An intriguing project in San Francisco aims to make something the participants call 'Real Vegan Cheese'. They modify ordinary baker's yeast so it produces the same fats and proteins as those found in milk, and then make cheese out of it. Though this isn't an unalloyed success so far, they've made a fair bit of progress. There are countless examples like these.

These community labs are linked with a fairly widespread, loose-knit movement called 'biohacking', or sometimes 'garage biology'. Biohackers are a motley crew of people trying to change their own bodies in various ways. The trend predates community labs: the term 'biohacking' was coined in the late 1980s, when changing your body in line with the same sort of principles you apply to writing a computer program became part of the broader hacker culture. Those in the mainstream focus on exercise, meditation, diet and dietary supplements. At the other extreme, there's a group sometimes known as 'grinders', who harness technology in an attempt to transform their bodies. This can range from implanting a radio-frequency identification (RFID) chip in your hand to inserting a device into your ankle that vibrates every time there's an earthquake somewhere in the world, like the Spanish dancer Moon Ribas. Moon channels this permanent connection with the seismic state of the planet into dance performances that interpret the earth's seismic activity in real time. Others seek to challenge the limits of human longevity, or even what it means to be human. The movement as a whole combines a playful approach with the will to exercise more control and stretch the limits of the human body, plus the belief that you can experiment on anything and everything.

Among these people, there's a group who are looking at the new genetic technology and thinking about how it can be applied to their own bodies. At the opposite end of the garage-biology spectrum from Ellen Jorgensen, we find Josiah Zayner. If she's the queen in the pack, he's the knave.

Blue-haired Zayner, who turned 40 in early 2021, once worked as a researcher at NASA. He left because he thought the agency too sluggish and rigid. In October 2017, at a conference streamed live on YouTube, he held forth – between slugs of Scotch – on the potential of gene technology. On the table in front of him lay a syringe. He launched into a long tirade about the need for a biological revolution and for this new technology to be made available to everyone, rather than being controlled by a tiny minority.

Then he injected himself in the forearm, becoming probably the first person ever to try to gene-edit himself

– quite independently, without the assistance of doctors or other scientists.[4]

If Zayner was telling the truth, the syringe contained everything needed to reprogramme some of the cells in his arm with the help of CRISPR. The idea was to switch off a gene that controls a growth hormone. Animals with a defect in that particular gene often experience massive muscle growth. 'It probably won't have much effect,' Zayner said straight after the injection. 'The amount of cells it's actually going to modify will be pretty small and it'll probably take a number of injections over a period of time to actually see a significant effect on my body.' The point wasn't to enhance his muscles, but to show that anyone could edit their own genes. This is now a fait accompli, and nobody knows how many others have tried out the technique by now. Netflix has since produced a series called *Unnatural Selection* about biohackers and the potential of the new gene technology, in which Zayner plays a major role.[5]

He's far from alone in experimenting with gene technology on his own body. Before him, Liz Parrish was probably the first to try out a genetic treatment which she'd developed within her own firm. She was injected with two different gene-therapy preparations which she claims will give her a longer and healthier life. No results have been published, nor is there even any detailed information about how the treatment was given. However, she now runs a company providing gene therapy that's supposedly designed to extend life expectancy. In June 2017, a doctor helped a 60-year-old American, Brian Hanley, to try to edit his own genes, to boost his strength and stamina. Little information is publicly available about that experiment either. Another example is Justin Atkin, a scientist who attempted to edit his own genes to rid himself of lactose intolerance and who has given an account of the results in an online forum.[6]

Whether such experiments are successful or not, they clearly involve a risk. That became abundantly clear in February 2018, when Aaron Traywick made his way onto the podium at a conference in Texas. He was the founder of Ascendance Biomedical, a company with its roots in the biohacking movement. Now, dropping his trousers, he pulled out a syringe and injected an experimental genetically modified herpes treatment into his left leg. Ascendance Biomedical hasn't released any documentation to say whether or not this treatment was successful, or even exactly how it was supposed to work. In the past, the company once got into hot water on account of a similar public experiment that was supposed to treat HIV with the help of CRISPR-Cas9. Moreover, a few months later Trayvick died, aged just 28. A forensic examination established that there was no link between his death and the injection.[7]

Since 2016 Josiah Zayner has been selling kits to people who want to try gene-editing at home. The standard kit is for modifying bacteria and yeast, but in 2018 he started selling a far more controversial product. US residents can order live frogs from his company, together with everything you need to modify their genome. The price is an affordable US$299, though the equipment may need to be augmented with the full genetic-engineering home-lab kit, which costs just under US$2,000. For an extra ten dollars, you can also buy crickets to feed the frogs. The gene you modify is the one controlling the animals' growth. If everything works, the gene-edited frogs grow to be much larger.[8]

Universities and companies have to obtain authorization and undergo regular checks to experiment on vertebrates (such as frogs), but so far the home kit has avoided regulation and legislation in the United States because the frogs are classed as domestic, not laboratory animals. Zayner himself says one of the aims of selling the kit is to make it easier for people to have a go at editing their own genes for the first time. The company's website includes a step-by-step guide on how to gene-edit yourself and how to order the chemicals and the genetic material you need. For those who really want to immerse themselves in the subject, Zayner also offers an online crash course in 'Advanced DNA Programming'. Everything can be ordered online, and it's all completely legal for the time being. It doesn't involve gene-editing embryos (germline editing), so there's no risk of the changes being passed on to successive generations. However, there is a pretty big risk that someone will harm themselves.

Zayner defends his projects with the argument that he's forever being contacted by sick people and their family members who want help with trying out treatments they can't access through normal medical channels. In January 2019 he wrote on his Facebook page:

> You know what's the worst? When someone emails me looking for help and they have an ultra rare disease that no one is working on treating. It is ILLEGAL for me to try and help them according to the FDA and US government. They are just supposed to suffer and die. Their only chance is someone like me willing to risk going to jail to help them. In our current system you need to be an outlaw to try and help stop someone's suffering. There's got to be a better way.[9]

He writes that he gets a lot of emails from people with various genetic muscular diseases. There's a fair amount of ongoing research into ways gene therapy can be used to help them. But most of those who are ill can't take part in new, experimental trials. 'Those emails are the hardest to ignore,' he writes, ending with the hashtag #PiratingGeneTherapy. He has also helped people develop their own treatments, though it's far from clear whether they will work.

At the same time, Zayner claims in interviews to be self-critical about the hype around people trying to hack their own genes: he's sure someone will come to harm. In other interviews, however, he's said that you probably run less risk of getting cancer by using CRISPR-Cas9 on yourself than by smoking, so why is the one thing more socially acceptable than the other? He clearly revels in his provocative stance.[10]

It's hard not to love a rebel. And Josiah Zayner fits the narrative of the hip young gunslinger who, tired of rigid bureaucracy, ventures out into the unknown to free the masses: a crusader for a new revolution in both biology and democracy. But Ellen Jorgensen doesn't buy that characterization.

'It's not just that they're aiming to get publicity, it's that they're aiming to get publicity to make money,' she says. 'So they have a financial interest in convincing people that what they're doing is cool and attractive and edgy, and in some cases I think it's highly irresponsible because they're not saying it outright but they're leading people to believe that their products can potentially be used to edit humans. And that's not true at all. It would take a lot of professional biotech expertise to be able to turn their products into anything that could have the slightest possibility of affecting a human.'

Most community labs are a far cry from that part of the biohacking movement. With their focus on educating the public, Ellen says, the last thing they want is to be conflated with the more provocative biohackers. She's clearly irritated by what she calls the 'attention-getting stunts' that have taken place recently. Another annoyance is the way community labs are portrayed in the media.

'It amazes me that we've been around for almost ten years and there are still people reporting on us as if they've just discovered us. It usually happens around the time that some sufficiently scary tech comes out and then it's "Oh my God, what if the DIY bio people get their hands on this tech?" And we've seen this cycle happen again and again and again,' she sighs.

When new gene technology becomes publicly available, it also becomes accessible to individuals and movements that don't have society's best interests at heart. The risk of bioterrorism gives great cause for concern. Not long ago, Canadian scientists published a paper on how to synthesize the horsepox virus, a relative of the eradicated smallpox virus. Smallpox is a terrible disease that has the potential to spread very rapidly. The scientists were strongly criticized for writing a paper that could be used as an instruction manual by terrorist organizations. There is a lively debate in scientific circles about what information, and how much of it, should be made public.[11] But Ellen Jorgensen thinks that community labs may well be able to play a role in combating bioterrorism.

'I would say that I have had conversations with pretty much all of the high-level people in homeland security in the United

States that are charged with keeping us safe from bioterrorism and none of them are worried about our community. Most of us are quite idealistic,' she says. 'Besides, it's very hard to do anything because everyone is always in everybody else's business. It's much more likely to happen in a professional or an academic lab. I know of at least two instances where people from my community alerted the FBI when they thought somebody in the lab was behaving a bit oddly. And nothing ever came of any of it, but here in New York we have the rule that if you see something, you say something. And if you're sophisticated enough to know what you're seeing, you're more likely to be able to say if there's something wrong.'

* * *

It's hardly new for people to worry about all the dreadful things that might happen if ordinary mortals were to start messing about in labs. A *Washington Post* article from 1988, 'Playing God in your basement', asked: 'Will beaming children, 10 or 15 years from now, be presenting genetically engineered sheep at the 4-H Club?'* The answer it gave was: 'No doubt about it.' The article makes interesting reading in that it shows just how much has changed, while the questions around those working with gene technology have stayed the same. It describes how gene technology is reaching new groups within society and points out the associated risks. Much of what is predicted to be possible 'in the future' is already with us: there are now kits that schoolchildren can use to make bacteria glow in the dark, and genetic tests to estimate the risk of particular diseases. Two sections at the end of the piece sum up exactly what remains worrying about 'bathtub biotech':

What happens, for example, if future generations begin to see life as something that's manipulable – just another computer program, but one in which the printout isn't on

---

* The 4-H Club is a network of youth organizations in the United States.

paper but in proteins? If children grow up believing that life is nothing more than organic chemistry? [...]

The truly frightening aspect of this technology isn't that the occasional outlaw will emerge. It's that society's beliefs about the nature of life will be so fragmented and confused that there will be no ethic for bio-hackers to emulate. In which case, all bets are off.[12]

Schoolchildren showing off their gene-edited goats is something that doesn't necessarily lie ten to 15 years in the future. In theory, it could happen tomorrow, if a school with extremely well-qualified teachers decided to make a major investment and to ignore the general public's views on animal ethics. The fact that it hasn't yet happened says a good deal about our society's capacity to take new technology on board without going to extremes. Concern about new technology undermining ethics and morality and changing our view of what is natural seems to have been a human trait for thousands of years. Despite this, we don't see too many secondary schools building missiles or nuclear power stations, even though the technology and the know-how are theoretically available.

More and more people are beginning to understand what can be achieved in community labs, while technology has become both cheaper and more accessible. Yet Ellen Jorgensen thinks the difficulty of applying gene technology is often underestimated:

'The analogy has been made quite deliberately that the biotech DIY community is quite like the Homebrew Computer Club. There's this expectation that it's going to be equally accessible to people when they enter the field. You can learn how to program if you just read enough books [...] but for biology you need a lot more equipment. You have to develop some hands-on skills. It's like me saying I want to be a five-star chef, so I'm just going to read a couple of books about it and walk into the kitchen and whip up this amazing meal. That's not how it works. There's all sorts of stuff you learn that can't be just out of a book, and the same is true for biology.'

Apart from all the practical know-how you need to make a biological experiment work, anyone who wants to edit their own genes in the basement will face the same problems that scientists do. There are some clear, simple changes that would make it possible to cure certain specific diseases or, for example, to temporarily change the colour of a few skin cells. But biohacking isn't like data hacking – for the simple reason that biology and computers aren't particularly similar, even if we happen to use the same terms in both fields.

The community labs and the biohackers who modify themselves are here to stay, for good or ill. They offer amazing opportunities – but they also involve the risk of people being led astray, or harming themselves. The perils go far beyond the risk of a horde of Hulk lookalikes flocking into the streets of New York. It may be that greater knowledge, particularly the type of practical know-how disseminated through the community labs, will provide the best kind of insurance against garage biohacking and gene technology going awry.[13] What worries today's scientists more is what happens when states choose to apply the new gene technology. And there's one country, above all, that's determined to get a head start on the rest of the world.

4

The Gene Race

On 4 October 1957, a small silvery sphere flew out into space. It looked like a shiny oversized beach ball with four long, slender antennae. In splendid isolation, it circled time after time around our planet, giving rise in equal measure to identity crises and burgeoning political strategies.

Sputnik 1 was the first satellite humankind had ever launched into orbit around the Earth, and anyone with an amateur radio set could pick up its regular beeps each time it passed. The fact that it was the Soviet Union that had launched it shocked the United States, undermining its self-image and setting off the space race. While the Soviet Union was the first country to send satellites, animals and people into space, the United States was the first to put a man on the moon. And the whole world benefited when the scientific breakthroughs and technical innovations that led to space travel began to be applied in everyday life – even though this progress was part of the Cold War.[1]

When Chinese scientists were the first to gene-edit human embryos in 2015, various international pundits started calling this a 'Sputnik event' – a signal to the rest of the world that China was flexing its scientific and technological muscles. The birth of the first gene-edited babies in 2018 fanned the flames.[2]

According to many business leaders, writers and intellectuals worldwide, we're now living in the century of biology. While the twentieth century saw a world revolution in physics, encompassing everything from electricity and the internal-combustion engine to nuclear weapons and space travel, biology and gene technology are set to trigger a revolution on an equal scale in the twenty-first. Just like the century of physics, this one will witness a tug of war over who gets to lead developments and pull the rest of the world into the new era – and, above all, who gets to decide what we're aiming for, what scientific questions we ask, what new 'Manhattan projects' are to be funded. At the moment, it looks as if China is determined to take on that role.

To catch at least a glimpse of this process, I head out to leafy Kunming in south-west China. By Chinese standards, Kunming is a fairly small city, with a population of just under 7 million. Its location on a high plateau makes for a pleasant climate all year round, giving rise to the epithet 'the City of Eternal Spring'. The area has been settled for over 2,000 years, and Kunming's green hills and forests are dotted with exquisite temples.

But outside the city centre, about an hour by car along the broad, multi-lane highways, an entirely new kind of settlement has sprung up, known as Science City. Lofty apartment blocks are scattered around like Lego bricks, in groups of ten, all just the same, all between 20 and 30 storeys high. In between them lie scientific institutes, businesses and research laboratories, surrounded by luxuriant vegetation. I make my way to one of the oldest buildings in this pioneering scientific city. It was built back in 2006, when the only things here were lush forests and fields.

China has done nothing to conceal its ambition to become the world's leading scientific nation by 2049, a century on from the proclamation of the People's Republic. Investing in science was important to the regime from the outset. Most of the country's inhabitants see China's rapid development, from a poor agricultural nation scarred by famine to the world's second-largest economy, as the result of the last few decades' big science-driven projects.

The government is currently investing billions of China's capital in building huge telescopes and countless research centres. A project called the 'Thousand Talents Plan' has been set up to try to tempt Chinese scientists who have built scientific careers abroad to return. The state is offering high salaries, extensive research facilities and brand-new laboratories to attract them.[3] One example is Pu Mu-Ming (known in the West as Mu-Ming Poo) at the Institute of Neuroscience in Shanghai, who, in 2018, became the first person to clone monkeys. When he set up the institute in 1999, he was the first research leader in China's Academy of Sciences to hold the citizenship of a foreign country. Though he was born in China, his family took him to Taiwan at the age of one when they fled from the Communist revolution. He went on to build a successful scientific career in the United States. However, a few years ago he returned to China to take up a leading post at the new neuroscience institute. He gave up his US citizenship in 2017 – a clear indication of how attractive China now is to scientists, regardless of where they stand politically.

China is a fascinating country with a sophisticated culture. But travelling around the country arouses considerable feelings of discomfort. A massive firewall blocks most Internet content. There are surveillance cameras everywhere you look. Bags are scanned, and you have to pass metal detectors every time you enter the metro or board a bus. Your ID card is checked constantly. The whole country is pervaded by a paranoid atmosphere. Twenty years ago, maybe even ten years ago – the last time I visited – China seemed to be on a slow, cautious trajectory towards a more democratic future. But it's now clear that the Chinese Communist Party holds the country in a vice-like grip.

I'm in Science City to visit the Yunnan Key Laboratory for Primate Biomedical Research, which specializes in gene-editing monkeys to make them easier to use in medical research.[4] On arrival, I'm given a guided tour by Xi Lin (or Amelie, her chosen Western name). Amelie says she thought about moving out here from Kunming, but decided against it. If she lived in Science City, her job would take over her life completely. So she spends two hours a day commuting by car. To kill time on her commute, she listens to the South African comedian Trevor Noah and plans her next journey abroad. When we meet, Thailand and Canada are high up on her wish list.

The research centre comprises several large buildings full of laboratories and offices, but I can hear the monkeys gibbering in the background. I'm not allowed to get very close to them, to walk among the cages, or to take any photos. This is partly to protect the animals from infection and disease; not even Amelie is allowed in among them. But the photography issue is also a matter of security and reputation. It's hard to get permission to photograph facilities with lab animals anywhere in the world, and pictures of baby monkeys behind bars are particularly poignant. So instead Amelie leads me up to a balcony, a purpose-built viewing point for visitors.

Behind a steel fence, there lies a field containing what look like small cabins, or perhaps unusually robust greenhouses. These are the monkeys' 'apartments', 76 in total, which together house nearly 4,000 macaques and rhesus monkeys. The enclosures are roofed over with transparent plastic and have open-latticework walls. Each 'apartment' contains a family group comprising a male, several females and a troop of youngsters. They have rings suspended from the roof, swings and rope ladders, both horizontal and vertical. There are small shelves jutting out from the walls, providing places to sit. I watch the monkeys moving around in their 'apartments', apparently relaxed and healthy. I'm no expert on animal care, and someone with more experience in that area might notice things I don't, but there's nothing in the monkeys' behaviour to suggest stress, lethargy or any other discomfort.

A few of the animals closest to us eye us curiously for a few seconds, before tiring; they can see us just as well as we can see them. A female perches on a rope ladder, swinging back and forth with a tiny infant in her arms.

'Mostly the monkeys can take care of their own babies,' says Amelie. But sometimes, if they're not up to it, or if their young have health problems, staff bring them into the lab instead. 'Like a human hospital, with the incubators.'

Other monkeys climb up the rope ladders or sit scratching themselves pensively. It's easy to get caught up in watching them, as if this were a zoo. Outside the cages, there are large nets which employees use to catch monkeys. The animals here haven't yet been used in any experiments. Amelie points out the kitchen, where they prepare the monkeys' meals, which are handed out three times a day and include fresh fruit and vegetables.

'Sometimes they even make cookies. It's like a small factory making cookies for the monkeys,' she jokes, and tells me the buildings are designed according to international guidelines: the institute is internationally certified. The monkeys have underfloor heating in winter and playthings to amuse them. There's nothing to suggest they might be suffering.

* * *

While these monkeys may not look particularly impressive, they're just one small facet of the tidal wave of Chinese science that's building up. When Sputnik was launched in the middle of the Cold War, the United States and the Soviet Union were openly at loggerheads. Now that China seems to have started a new science race, it isn't clear who its rivals are. Is China's investment in gene technology something the rest of the world should worry about? A leader in *The Economist* summarizes the situation rather neatly:

> The looming prospect of a dominant, rule-breaking, high-tech China alarms Western politicians, and not just because of the new weaponry it will develop. Authoritarian governments have a history of using science to oppress their own people. China already deploys AI techniques like facial recognition to monitor its population in real time. The outside world might find a China dabbling in genetic enhancement, autonomous AIs or geoengineering extremely frightening.

Results and discoveries from China's scientists will benefit the world as a whole. But China's research, scientific collaboration and invitations to visiting international scientists to use its laboratories, telescopes and particle physics installations are also an element of the country's soft power, one of the many tools it can exploit to get its way. They are also a tool for controlling the people of China themselves.

Gene technology has acquired a special place in China's scientific endeavours. This is such a new field that there is a great deal to be discovered and relatively little competition for the time being.

Amelie and I go up to talk to Professor Ji Weizhi, who founded the centre over a decade ago. Age notwithstanding – he's in his sixties – he is the scientific powerhouse behind the institution's new scientific breakthroughs. Professor Ji's English is slightly halting, even after several years at various universities in the United States. He belongs to the first generation that built science in China after the Cultural Revolution, when the country was gradually opening up to the West. A tall man with a pleasant manner, he is intense, clear and to the point. In his view, there is no doubt that CRISPR will be used to genetically modify human beings.

'I hope we can apply gene technology to make people healthier in future. We can avoid complex diseases even before birth. Some people carry mutations in their genes that cause disease, and that causes a lot of problems for families and society. If gene technology can solve this problem, I think it should be used to make our society healthier. So that is my hope.'

Forty years ago, when he began his scientific career, China was still a poor country with few scientific resources. He began with research on monkeys and became involved in conservation work to protect China's threatened primates. Many years later, he started to develop an interest in human health and to think about how he could contribute thanks to his knowledge of monkeys.

'We humans have a lot of problems with our health, especially with complex diseases,' he says. 'So I thought maybe we could establish some monkey models for human diseases. Maybe that would be a good way to solve this problem.'

Using animals as models, templates for human diseases, is a well-established practice. The underlying idea is that you can acquire a deeper understanding of what goes on in the human body by examining how a disease develops in animals. With laboratory animals, scientists can take samples in the course of a disease or trial new medicines in ways that would not be possible with human subjects. The problem, however, is the many diseases that afflict us, but not other animals. Ji Weizhi and other scientists in the field have made significant breakthroughs in this area.

'In 2014 we published the first paper on using gene-editing to establish a monkey model for a human disease,' he says proudly.

The first monkeys to be born were Ningning and Mingming, two macaque sisters who both bore modifications in two genes that affect metabolism and the immune system.[5]

'They are still alive and healthy,' says Ji Weizhi.

The monkeys were modified when the embryos still consisted of just one cell, so that the adult animals now have modifications in every part of their bodies. They are not the first ever genetically modified monkeys, but the CRISPR-Cas9 method enables scientists to make simpler and more precise changes, and at a lower financial cost. Efficiency and precision are a significant condition if the monkeys engineered in this way are to help form the basis for new scientific breakthroughs. This is yet another of the many groundbreaking studies conducted by Chinese scientists in recent years.[6]

'After the Chinese economy started to grow and the government started to support scientific research, the amount of funding increased,' says Ji Weizhi. He thinks they should be aiming to reach the leading edge of fundamental scientific research.

In 2018 China became the country that produces more scientific articles than any other in the world. It now spends more on research than any other country.

Large numbers of gene-edited monkeys have been born in this centre over the last few years. One of the diseases they are working on is Parkinson's. Ji Weizhi tells me so much time has passed since the first births that they have begun to observe differences in the brains of the genetically modified monkeys as they age.

'This is a disease that occurs with ageing, but now we've had a few years to observe these monkeys. And we've found that the brains of the gene-edited monkeys are becoming abnormal as they age.'

They haven't published their results yet and want to collect more data, so Ji Weizhi is circumspect about giving details during our conversation. His expression reminds me of a cat that has discovered an extra-large bowl of cream, so there may be exciting news on the way. He thinks their study is likely to be the first to show how the brains of monkeys with Parkinson's develop, a result that could be extremely important for our understanding of how diseases develop in human beings.

A better model of Parkinson's would also open up the possibility of developing new medicines. Parkinson's is a complex disease: it differs from many others in that it is caused by a combination of various genes and other factors. The models most often used by scientists today are mice and rats, but they are so different from us in both lifespan and brain structure that the results obtained are not always of any use.[7] Scientists have conducted experiments on primates using a variety of drugs and toxins to create a condition similar to Parkinson's. But Ji Weizhi is sceptical about the usefulness of such experiments. The disease isn't caused by such substances in humans, and we don't yet know exactly what happens in the brain, he explains.

Ji Weizhi and his colleagues are working to develop models for many diseases, with a particular focus on neurological disorders. As well as Parkinson's disease, they are researching motor neurone disease (amyotrophic lateral sclerosis, ALS), Duchenne muscular dystrophy (DMD) and Rett syndrome. Within three to five years, he believes the centre will have created a monkey that is good enough – that is, sufficiently similar to humans – for it to be used in trials of new medicines.

He also believes it will be possible to create models, genetically modified monkeys, for disorders such as autism, dementia, schizophrenia, and the like. 'The competition is tough in the world of science, so we have to work intensively,' he replies when I observe that these aims are very ambitious and the time frames for achieving them pretty tight.

Afterwards, once I've finished talking to the professor, Amelie takes me over to the other part of the facility, to the lab housing the monkeys currently being used in experiments in the institute. This is a far cry from a friendly zoo. I'm not even allowed to see the monkeys. There are plausible safety reasons for this. I don't want to ruin an experiment costing hundreds of thousands of dollars by accidentally infecting a monkey with a Swedish cold virus. But I suspect there's another reason: that these monkeys have a far less cosseted existence.

Instead, I'm allowed into the monitoring room. This has two large screens, one showing data from the lab – temperature, atmospheric moisture, and so on – and the other displaying images from surveillance cameras placed all over the lab.

The monkeys I see on the screens don't look nearly as relaxed and healthy as the ones outside. They are housed in small steel cages, and many are circling round and round or sitting still, with a fixed stare. When the camera moves, the monkeys follow it with their eyes. They open their mouths and gape at it. They may be shrieking – there's no sound, so it's hard to tell – or they may just be yawning. Their behaviour suggests both stress and boredom, but here I have to repeat that I'm no expert in this field and can only give my impressions. I'm sure an expert in animal care would have noticed other details.

The cages measure 80 by 80 centimetres. 'American standard,' says Amelie. Each houses one or two monkeys, and the cages are piled up in layers and lined up along the walls. There are about a thousand animals here in total at any given moment. Some of them are here for a brief period only, others for years. Some of the monkeys have a small, brightly coloured toy in their cages to occupy them. Amelie tells me that their keepers swap these playthings around regularly so that the animals get to see something new from time to time.

One monkey we observe for a while has realized that it can stick its hand out through the hole where its food is put into the cage and reach the locking mechanism. It sits fiddling with the lock during the minutes the camera follows it.

I get to see Ningning and Mingming – the first two monkeys to be genetically modified using CRISPR-Cas9 – on camera, in the same sort of cages as all the others. They're four years old now.

I am told that a scientist from Stanford is to arrive here next month to work with the institute, which is also collaborating with scientists from Stockholm's Karolinska Institute on a study using genetically modified monkeys to research heart diseases. Over the last few years, many laboratories like this one have appeared in China. At the same time, many comparable labs have closed down in the United States and Europe. More and more American and European scientists are choosing to move their experiments on primates to China. Research using monkeys as lab animals is in the process of shifting from one continent to another.

There are no exact figures on the number of monkeys used in experiments in China, but all observers are agreed that the number has shot up over the last few years. A report from 2008 estimated that there were 40 laboratories in China with over 40,000 rhesus monkeys and 170,000 macaques. There are certainly more such labs and more monkeys today. The monkeys are exported to various countries or used for experiments within China.[8]

Before scientists in the United States or Europe are authorized to experiment on primates, their plans must be scrutinized by an ethics committee. This can take months. In China, the process is far more permissive and relaxed, and sometimes even non-existent. It is likely that monkeys are used in studies that wouldn't have been permitted in other parts of the world. But another important reason for the many primate facilities opening in China and the fact that scientists from the rest of the world are moving their research to this country is that it is significantly cheaper to breed monkeys here. A third reason is that there is essentially no resistance by animal-rights activists or others to this kind of facility in China. Amelie says they haven't experienced any threats or protests.

'Up to now, no. Not yet. But we are very cautious about that. First we keep it low-key, and second we try our best to keep the animals healthy, alive and happy.'

In Europe, all scientists working with lab animals have to contend with personal threats and the need for security equipment around the buildings and other facilities – irrespective of the kind of animals they work with. I ask Ji Weizhi about his views on animal experiments.

'I believe a lot of people worry about ethics, but [...] to use monkeys [...] we have to follow some regulations, such as the three Rs: reduction, replacement and refinement. This means, if we can use another animal or test, we do not use monkeys. [...] And also we need to design our experiments to be efficient, not to waste animals.' He thinks we Westerners are over-sensitive: 'I believe in the Western world people are too focused on ethics. That's not so good for science. In China, things are much better than in the Western world: the people that worry about ethics are concerned about their pets,' he says.

Ji Weizhi finds it a little difficult to understand the Western view of ethics, and why so many people are critical of experiments on primates. Can I explain that to him? I have to think quickly. 'People in the West are mostly healthy and live long lives. I think that might be what makes them more critical of this kind of research,' I say.

Once I'm back in the hotel and have enough time to reflect, I realize that my reply covered part of the explanation, but not all of it.

We continue to succumb to disease, to complex illnesses with complicated and diffuse causes. A toxic brew of genes and environment gives us cardiovascular diseases, cancer and the disorders that cause us most discomfort. Just like most others, I know people with Parkinson's, Alzheimer's and dementia, diseases that cause terrible suffering. Yet there's hope that these diseases can be alleviated, and perhaps even cured, through the very research I'm looking at here. And this hope is stronger now that we can edit monkeys' genes so that they get the same diseases as we do.

People in the West have the resources and energy to criticize animal experiments. We can extend our empathy beyond our own species: our prosperity gives us the security to concern ourselves with other creatures' needs. But how is one set of needs to be weighed against another? How much suffering is it reasonable to inflict on an animal in the hope of relieving the suffering of a human being?

Ji Weizhi doesn't look at these issues in the same way. He thinks it's clear that animal experiments are the right way to go. I ask him what risks he associates with gene technology: what impact could it have on society?

'I think every new technology is something that people worry about. But I don't think it's really a problem,' he says. 'It's like transgenic food a few years ago. So far I think it is a good technology. And it can improve human lives, especially in developing countries. So gene-editing, I believe, after scientists have really deeply studied this technology and its safety aspects, will be very useful in the future, for society and for human beings.'

During my conversation with Ji Weizhi, I ask him about developments in China and where the country is heading in research terms.

'Within the next ten years, maybe we can compare with the US or European countries. Will we be better? I don't know. It's difficult to say; it depends on the next ten years in the policies and economies of the different countries.'

Despite the huge investments in Chinese research, science isn't free in the same way it is in a democracy. A number of studies and interviews of scientists working in China highlight the problem of the lack of scientific freedom, which follows on from the lack of civil and political freedom.[9] The Party's decisions and views are paramount at all times, regardless of any other factors. Yet for research to be effective, and for the country to attain the results the government wants, much more freedom is presumably needed. It's hard to make any kind of breakthrough if you can't question orders or raise objections to old ways of thinking, or if you are used to constantly censoring your thoughts and taking care not to challenge paradigms and hierarchies. There are also reports indicating that even Chinese exchange students abroad are under pressure to continue abiding by the regime's wishes.

Scientists develop by sharing ideas with colleagues, regardless of where in the world they work. How China is to resolve that contradiction remains a mystery. Moreover, history shows a link between research and the fight for political freedom.[10]

During my conversation with Ji Weizhi I avoid mentioning the fact that it is the government, above all, that concerns people both within and outside China. This isn't a question that produces replies journalists can use, as I've noticed during my travels. Instead, I ask if there's anything he'd like to say to those concerned about the rapid development of gene technology.

'I don't think they should worry about this,' he says. 'Scientists have their honour. I think most of them will do the right thing: they don't want to do evil things. Most of them, maybe 99 per cent of them, are nice guys. Good men. So don't worry. And I think in the future scientists will resolve the problems of these techniques and improve on the limitations. We just want to give human beings a very happy and healthy life.'

* * *

In the course of our meeting, Ji Weizhi strikes me as being an honest person – an impression that changes a few months after I have returned home from my travels.

During the desk research I carried out before visiting China, I came across the name of another scientist at the institute where Ji Weizhi works – Su Bing. He had said in a number of interviews that he aimed to attempt the genetic modification of monkeys' brains to make them more like those of humans. His aim was to splice in human genes to see whether the monkeys became more intelligent or whether it was possible to elicit uniquely human characteristics. Su Bing leaves my emails unanswered when I try to contact him, so I ask Ji Weizhi about him and his experiments.

Ji Weizhi strongly denies involvement in any such experiments, claiming that Su Bing has left the centre and that they no longer work together.

But six months later Su Bing publishes his results in a Chinese scientific journal.[11] The experiments were conducted at the same centre, and Ji Weizhi is listed as one of the co-authors. The scientists took a gene that appears to be linked to the size of the brain in humans and inserted it into a few monkeys. After that they measured the monkeys' brains and carried out various cognitive tests. I will come back to their results in Chapter 12. These experiments must have been going on at the centre while I was visiting. I may even have seen some of the animals involved.

These experiments are regarded as deeply unethical by most of the world's scientists, and were strongly criticized after the results were published. They are a world away from the ethical guidelines 'reduce, replace and refine', which imply conducting as few animal experiments as possible and carrying the tests out as well as possible, and only in studies whose results may have a significant impact in relieving human suffering.

Reading the article, I reflect on Ji Weizhi's words about honour. But to really understand where developments in China and the rest of the world can lead, we need to take a rather closer look at how gene-editing actually works.

5

Storm in a Bowl of Yoghurt

Certain topics divide humanity into two camps, sowing discord and arousing feelings so strong that it's almost impossible for one side to understand the other. These are subjects you avoid talking about at family dinners or during office tea breaks, so as not to put a damper on things. One such *casus belli* is the world of microorganisms – bacteria, viruses and other single-celled life forms, like mould and amoebas.[1] Much of the world's population is revolted by mouldy bread, rotting fruit, babies' nappies and the runny noses that go with the common cold. They'd rather not think about what's behind the unpleasant smell. And then there are the rest of us, mostly biologists – and a few other courageous souls – who observe the same phenomena in fascination, spellbound by an ecosystem where conflicts and cooperation play as important a role as in the African savannah.

This chapter takes a deep dive into the world of microbes, with a focus on gene technology and the science and research revolution set off by CRISPR-Cas9, the gene scissors. If you're just interested in how this technology affects people and society in general, you can skip this chapter. The next one, in which I explain what exactly we mean by genetics and genes, also zeroes in on the science. But if you're brave you'll follow me on my journey into this microscopic realm, which offers a world of fascination.

Just as in the natural surroundings we're familiar with, unrelenting rivalry pervades the ruthless world of microorganisms. Only the fittest survive. Each minuscule advantage is exploited relentlessly, so that both attacks and defences grow ever more sophisticated. One of the best-known examples of the combatants on this microscopic battlefield was discovered by pure chance.

The Scottish scientist Alexander Fleming had a keen interest in bacteria. In the summer of 1928 he happened to be studying a group known as staphylococci, which, though often harmless, can sometimes cause infections. He cultivated them in Petri dishes containing agar jelly, a growth medium, as is still the practice in today's laboratories. Fleming took a long holiday with his family in August, but neglected to tidy up his experiments before leaving. Rather, he just piled up the Petri dishes with their bacterial cultures and left them in a corner. He already had a reputation as a brilliant scientist with a rather devil-may-care attitude to clearing up. On his return in early September, he observed that one of his Petri dishes was no longer usable, as a mould had invaded the bacterial culture. These things happen, but the odd thing was that the bacteria surrounding the mould had died, while other bacteria in the dish continued to flourish. The bacteria's inexplicable death was a sufficiently tempting mystery for Fleming to start cultivating the same mould on culture plates and observing what happened if he combined it with different strains of bacteria. The result was replicated again and again: the mould released a substance that killed nearby bacteria. To begin with he called it 'mould juice', but in March 1929 he announced to the world that he had discovered penicillin. 'One sometimes finds what one is not looking for. When I woke up just after dawn on September 28, 1928, I certainly didn't plan to revolutionize all medicine by discovering the world's first antibiotic, or bacteria killer. But I suppose that was exactly what I did.'[2]

Since then, scientists have discovered many other types of antibiotics, all of which come from the defensive systems used by microorganisms to combat bacteria, and they have transformed both medical care and society as a whole. The first real breakthrough was in care for the wounded during the Second World War, since when antibiotics have been put to a multitude of uses: from operations to the treatment of pneumonia; from curing gonorrhoea and Lyme disease to treating recipients of donor organs; from treating childbirth-related infections to fighting syphilis and tuberculosis (TB).[3] Today we also use antibiotics for less positive purposes: they are fed to poultry, cattle and pigs to promote faster growth and avoid disease in crowded spaces. They are also sprinkled over farmed salmon. There was even a period in the 1950s when meat was dipped in an antibiotic solution so it would keep longer.[4] All in all, it's nearly impossible to imagine what the world was like before antibiotics. However, since we started to deploy these substances, bacteria have also entered the arms race, finding new ways to shield themselves against antibiotics. They have developed resistance: more and more of them can now resist our new weapons faster than we can develop new ones.

The discovery of antibiotics is by no means the first time we humans have appropriated microbes' creativity for our own purposes. In breaking down sugar to form alcohol, yeast also creates an environment that's inimical to bacteria. And the bacteria that produce lactic acid in sauerkraut and fermented milk keep yeast-based moulds at bay and prevent food from rotting. Just a few years ago, there was a new breakthrough, a new way to turn the battle between microorganisms to our own advantage.

* * *

It began with an unexpected revelation. In the 1980s, Japanese geneticists discovered a curious phenomenon in bacterial DNA: minute snippets of genetic code that looked the same regardless of whether you read them from front to back or back to front – like the words 'kayak', 'level' or 'madam'. A feature of these mini-palindromes was that they seemed to occur in clusters. The scientists noted the discovery but didn't pursue it any further. It was just another of nature's many curiosities.[5]

These odd bits of genetic code continued to make an appearance in scientific research into bacteria. In the early 2000s it became known that the phenomenon had some kind of link with bacteria's immune system. By that time, the curiosity had also been given a name – CRISPR. The acronym stands for 'clustered regularly interspaced short palindromic repeats', a name that tells us little about what these 'repeats' actually do: an eloquent illustration of how scientists give little thought to marketing or communication when they name phenomena.

Viruses are another of the big players in the world of microbes. The word 'virus' now brings COVID-19 to mind – or possibly bird flu or Ebola. Yet the vast majority of viruses don't attack us humans, or even other animals: the largest group of them, phages, attack bacteria. Wherever bacteria are to be found, which is pretty much everywhere, you'll also find viruses attacking them. According to some calculations, these particular viruses actually outnumber all other organisms put together, including bacteria.

The way viruses spread is almost diabolically crafty. A virus is just a capsule, rather like a tiny spacecraft, that cannot grow or reproduce independently.[6] Instead, it attacks a cell – such as a hapless bacterium – alighting like a fighter plane on its surface. It then drills its way through the outer cell wall and squirts its contents into the bacterium. Most of what it injects is the virus's genome, which now floats around within the bacterium. This genome takes over the internal system of the bacterial cell, forcing it to make new copies of the virus which can then spread to new victims, after which the whole process is repeated. It's like a cross between a fledgling cuckoo and a malevolent photocopier. Sometimes the viral genome seeks out the bacterial genome; it then breaks in and infiltrates it, after which it can remain cocooned there for a long time.

Yet bacteria are far from being defenceless organisms. Many have developed a set of specialized weapons to fend off viral infiltrators.[7] The curious CRISPR code is linked with one of these.

The latest breakthrough took place in a bowl of yoghurt. Firms producing yoghurt, cheese and other foods derived from the action of bacteria on milk face sporadic problems with viruses that destroy lactic acid bacteria, spoiling the product. But scientists linked with the company Danisco discovered that bacteria whose genome contained a lot of CRISPR sequences were better protected. They realized that these chunks of DNA were a kind of reference library: the bacteria collected and saved short sections of genetic material from different kinds of virus to compare them with future viral invasions, enabling them to protect themselves.

Shortly after that, the scientists found out how these bacteria deployed the CRISPR sequences to combat viruses. It turned out that there was an enzyme (a protein) which, having received instructions from the CRISPR 'library', moved around inside the cell. It identified intrusive viruses, attached itself to them and cut up their genomes – like a pair of scissors homing in on their target. Hence the nickname often given to CRISPR – 'gene scissors'.

CRISPR systems often work like self-destructing weapons inside bacteria, killing both the virus and the bacterium. Sometimes there's a straight, clean cut, and sometimes the DNA molecule is chewed to bits.[8]

Up to this point, CRISPR might have been just one more interesting, rather curious biological function, another of the intriguing features developed in the arms race between different kinds of microbes. But two scientists – Jennifer Doudna, at the University of California in Berkeley, and Emmanuelle Charpentier, who was researching at Umeå University (Sweden) at the time – succeeded in showing that one of the many CRISPR systems could be turned into an elegant and practical tool. A tool so revolutionary that the two were awarded the Nobel Prize in Chemistry just a few years later, in 2020.

There are many variants of the CRISPR system, but the one that's taken the world by storm is called Cas9, so the whole package is sometimes referred to as 'CRISPR-Cas9'. 'Cas' stands for 'CRISPR-associated', while the number 9 indicates the precise type of Cas enzyme involved. CRISPR-Cas9 is a bit of a mouthful, but we'd better get used to it. People probably thought 'electroencephalogram' and 'amoxycillin' equally clunky when they first appeared.

* * *

Jennifer Doudna and Emmanuelle Charpentier had begun their collaboration a year previously, after meeting at a conference. Doudna was investigating bacteria with particularly large numbers of CRISPR sequences, and Charpentier had already played an important role in establishing how CRISPR worked. So it made sense for them to start working together. Doudna has said in interviews that she thought their project would probably be the most obscure part of her whole scientific career, but what happened was quite the reverse: it resulted in a huge breakthrough. The pair were the first to show that CRISPR could be used to modify genes in bacteria. Their method of genetic modification is simpler, cheaper and far more elegant and precise than any discovered previously. They opened up a whole new scientific world. 'Practically overnight, we have found ourselves on the cusp of a new age in genetic engineering and biological mastery – a revolutionary era in which the possibilities are limited only by our collective imagination,' writes Doudna in her book about the discovery of CRISPR-Cas9, *A Crack in Creation* (2017).[9]

This is where the discovery starts to resemble Alexander Fleming's breakthrough with antibiotics: it's a completely new way to use a biological system. Several research teams published their results at very short intervals just a few months after the first study came out. Feng Zhang at MIT and Harvard's George Church were the first to show, in 2013, that CRISPR-Cas9 worked in animal and human cells. The Lithuanian scientist Virginijus Šikšnys discovered a similar

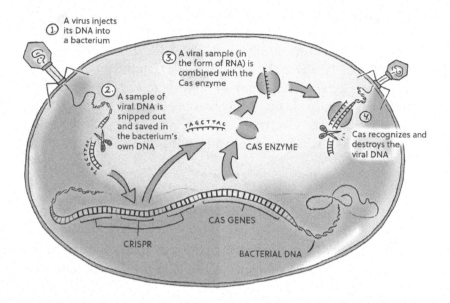

The CRISPR system was developed by bacteria to protect them against viruses. The Cas enzyme uses the CRISPR 'library' to locate and destroy the virus's genetic material. There are many different CRISPR systems in nature, all of which work in different ways. RNA is a variant of DNA which is formed when chunks of genetic material are copied for use within the cell, by the Cas enzyme, for example.

solution at almost the same time as Jennifer Doudna and Emmanuelle Charpentier. A multitude of other scientists added thousands of small steps to the rapid development we are now seeing. This technology has led to hundreds of patents and patent applications and thousands of scientific studies. Scientists working in many laboratories are eagerly investigating other CRISPR systems in bacteria to see whether they can discover even better tools somewhere in the bacteria's defensive systems. The reason why CRISPR has become so important so fast is this: the gene scissors can receive instructions and cut through almost any kind of DNA – including that of humans and animals. The scientists had discovered a biological universal solution, an adjustable spanner that fits any nut you care to choose.

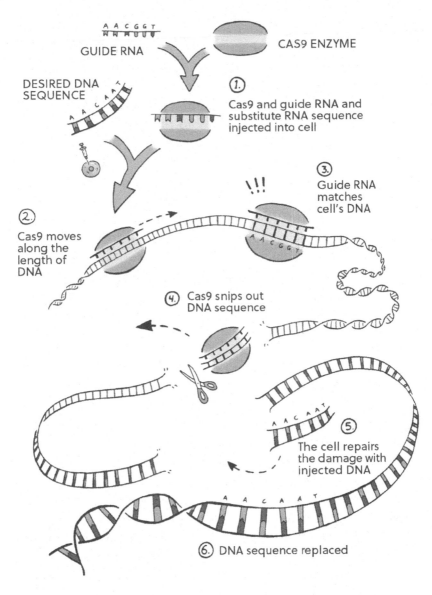

AACGGT
GUIDE RNA

CAS9 ENZYME

DESIRED DNA
SEQUENCE

1.
Cas9 and guide RNA and
substitute RNA sequence
injected into cell

3.
Guide RNA
matches
cell's DNA

!!!

2.
Cas9 moves
along the
length of
DNA

4. Cas9 snips out
DNA sequence

5.
The cell repairs
the damage with
injected DNA

6. DNA sequence replaced

Scientists can direct CRISPR-Cas9 so it modifies the genetic material inside cells. The Cas enzyme uses a tiny chunk of genetic material as a guide, compares it with the string of DNA and cuts out a matching DNA sequence. The cell can either insert a new chunk of genetic material or simply remove parts of the length of DNA. RNA is a variant of DNA that forms when chunks of genetic material are copied for use within the cell, for instance so that the Cas enzyme can use it.

* * *

All the genetically modified organisms predating 2012 were produced using methods that were costly, slow and extremely inefficient. It was hard to determine the location where a new gene was to be spliced in and incredibly awkward to modify more than one gene at a time. The process can be compared with trying to hit the bull's eye with a shotgun. Although a piece of shot hits the mark once in a while, most of the shot misses, and much of it rips the target apart in the process. The CRISPR method has radically changed all of this, being cheaper, simpler and faster. Scientists have become sharpshooters overnight.[10]

This technology is so cheap that it has already spread to labs worldwide, even in the world's poorest countries. It's so simple to use that members of the public can now order kits for home use. Above all, it's so fast and efficient that scientists can now dream up entirely new things to do with genes. And all this has come from free-riding on bacteria's antiviral system.

While the system used by bacteria provides the basic model, in the few years this technology has existed scientists have already developed several variants enabling the method to be put to a variety of uses. The simplest application is removing or dismantling a gene, such as a variant resulting from a harmful mutation that causes a hereditary disorder. The scissors are instructed to locate that gene, zero in on it and cut out the unwanted genetic material, after which the cell's repair system can mend the gap. Another option is to use the system to splice in a gene from somewhere else: one that enables yeast cells to produce insulin, for instance, or a substitute for a defective human gene.

The most recent versions of this technology are so refined that, rather than removing and inserting large chunks of genetic material, scientists can now alter individual letters in the genetic code, rather like correcting spelling mistakes. CRISPR can also be used to check whether particular genes are present, without damaging them, and to improve our understanding of how individual genes function by switching them on and off. This may be a way to improve our understanding of diseases – or to eradicate them for good.

* * *

Scissors are an apt metaphor for CRISPR-Cas9 because the system cuts through genetic material. CRISPR-Cas9 has also been compared to a Swiss army knife because it allows you to do various things after cleaving a chunk of genetic material: you can excise it, add more material or modify what's there. A third analogy is the word processor: CRISPR lets you rewrite a genetic code as if you were sitting at a keyboard. Many scientists now speak of gene-editing, rather than genetic modification.

Each of these three analogies – scissors, Swiss army knife and word processor – capture a facet of how the system works, making it a little easier to grasp what happens at cellular level. But to find an analogy that captures the resultant societal change more aptly, we need to go back to the fifteenth century. In 1439 Johannes Gutenberg invented the modern European printing press. This allowed every letter on a page to be moved around and reused for new pages. A similar system had been in use in China since 1041, but since there are so many more Chinese characters than there are letters in the Latin alphabet, the Chinese printing press had less impact; it wasn't really any handier to use than other methods.[11]

In Europe, however, the printing press set off a revolution: suddenly it was possible to make books that were easy to read, attractive, yet relatively inexpensive. Once texts could be rapidly disseminated, it became possible to print newspapers, magazines, pamphlets, posters and all kinds of other media that would once have been expensive and hard to produce.[12]

In the longer term, the more affordable texts which printing made available boosted literacy. It became easier to spread ideas of various kinds, which in its turn sparked off numerous social changes great and small – from the scientific revolution, which enabled scientists to disseminate their results with ease, to the organization of workers in trade unions, which established the necessary conditions for a more equal society with more rights and higher wages. The printing press paved the way for countless small-scale phenomena that together changed society as a whole.

Within just a few years, CRISPR and the new gene technology have set off a similar revolution. It began in research laboratories, but it's now well on the way to spreading out into society. Change is under way, but the huge impact it will have has yet to attract widespread attention. At the end of the fifteenth century, it would have been hard to imagine the cascading developments that would result from the printing press, developments that had already begun to reshape the world. We're at about the same stage with CRISPR now.

In addition, there are two other new methods for genetic modification of cells that appeared shortly before CRISPR. Just like CRISPR-Cas9, TALEN and zinc fingers enable scientists to identify and control the exact location where a genetic modification is to be effected. These methods are more costly and tricky to use, but experience with their use paved the way for CRISPR, and there are cases in which they work better. After the discovery of the CRISPR method, scientists have also continued to experiment with variants based on the immune system of a bacterium, which they manipulate in order to maximize the precision of genetic modifications. These are incredibly important discoveries, but they haven't garnered as much attention as CRISPR.[13]

This method has already been applied to everything from bacteria to human beings and has inspired great optimism in the scientific world. Will there be something new after CRISPR? No doubt there will, but it will only speed up a process that is already in motion – just as typewriters and tablets followed on from the printing press. The changes that the new technique has made possible will only accelerate as more discoveries are made and the methods in use are refined.

CRISPR and the new gene-technology revolution have other points in common with antibiotics. Just like Alexander

Fleming, the CRISPR scientists discovered that we humans can benefit from a phenomenon originally developed as a weapon in the ongoing war between microorganisms. Moreover, both CRISPR and antibiotics were initially the result of chance, discovered by scientists who were concerned more with understanding the world than with developing new tools. After all, it's impossible to deliberately seek out so revolutionary a phenomenon.

The discovery of the gene scissors may have an even greater impact on humanity than the Petri dishes Alexander Fleming left lying around in his lab. But while we now have a revolutionary tool at our disposal, we still need to understand what exactly it is we want to modify.

6

Unruly Peas

In the mid nineteenth century, a friar living in what is now Czechia (then part of the Austro-Hungarian Empire) started cultivating peas. There's nothing out of the ordinary in that, of course: most monasteries, then as now, grew some of their food in their own kitchen gardens. This particular friar, later to become abbot of the Abbey of St Thomas in Brno, experimented with crossing different pea varieties. There was nothing especially startling in that, either, even if he did have an unusually large garden at his disposal. For ever since humans started cultivating plants or rearing animals – over 12,000 years ago, that is – we've also done our best to maximize desirable features through selective breeding and crossing. We have created thousands of breeds of cattle, sheep, chickens and other animals, and even more varieties of fruit and vegetables. Farmers have always known that calves from a particular cow, or lambs sired by a particular ram, are slightly bigger or better or hardier than the offspring of other cows or rams. They've always known that mating a particular large bull with a cow that gives a lot of milk sometimes produces a big cow with a high milk yield, and sometimes a feeble little cow that gives no milk at all. We came to understand the workings of breeding and heredity through experience, almost by instinct. The human race has been able to exploit that knowledge to change nearly all the animals and plants around us, be they dogs or kumquats.

But what set our friar apart from most gardeners before him was that he started to make calculations. Some of his pea plants' characteristics seemed to be either one thing or the other: for instance, the peas were either yellow or green, but not a yellowish green. If he crossed a plant bearing yellow peas with one yielding green peas, all the peas in the first generation would be yellow, but if he then crossed two of those plants with each other, three quarters of the resultant peas would be yellow and one quarter green. Crossing plants with green peas always produced green peas.

If this is beginning to sound familiar, it's because most of us have studied these peas and their various combinations at secondary school, in our biology classes on heredity. The friar, Gregor Mendel, and his experiments with 28,000 pea plants in the abbey's experimental garden laid the foundations for what we today call genetics.[1] Mendel was the first to show that there are traits which are passed on in a pure, discrete form from parents to offspring.[2] He referred to invisible 'factors', while today we use the word 'genes'. He also showed that some of these factors could override others.

The classic example familiar to us from our secondary-school lessons is eye colour. Whether our eyes are blue or brown depends almost entirely on a single gene – a factor, as Mendel would have called it. One variant of the gene codes for brown eyes, another for blue eyes. This gene is located on chromosome 15, and since you have two chromosomes – one from your mother and one from your father – you also have two of these genes. If you inherited a gene for brown eyes from your mother and a gene for blue eyes from your father, your eyes will probably be brown. This is because the gene for brown eyes is dominant, that is, it overrides the other gene. But if you procreate with someone who has the same genetic makeup as you in this respect – a 'blue gene' and a 'brown gene' – the two of you could have both blue-eyed and brown-eyed children. If you had four children together and each of them inherited one of the four genetic combinations, one of them would be blue-eyed – the child who inherited the gene for blue eyes from both of you. Two would be brown-eyed, but would carry a gene for blue eyes, just like their parents. The fourth would also have brown eyes, with a copy of the gene for brown eyes from each parent.

It's simple and straightforward to work this out and demonstrate it on paper. Gregor Mendel studied seven separate traits in his peas and showed that all of them conformed to the same pattern. Discrete traits, dominant and recessive factors. There's something incredibly satisfying about experiments and theories of such simplicity and elegance, patterns that make the world fall neatly and comprehensibly into place.

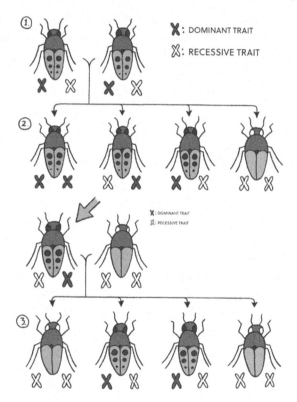

A diagram showing crossing between two beetles. Dark eyes and a spotted shell are dominant traits. It's enough for one of the parents to have dark eyes for their offspring to have dark eyes as well. For a young beetle to have light-coloured eyes and a plain shell, it must inherit the recessive trait from both parents.

However, it must be said that the world tends not always to conform to elegant explanations of this kind. In this particular case, its nonconformity mostly takes the form of a teenage pupil asking: 'But what about people with green eyes? Or people with one brown eye and one blue eye? Or people whose eyes are dark grey, not blue? And brown eyes can be dark brown or light brown, can't they?'

Gregor Mendel also encountered the same problem: peas that didn't behave quite as they were supposed to, percentages that didn't fit, plants that were generally nonconformist and refused to play by the rules. We don't know exactly what he did, but later scientists who examined his results swiftly realized that they are a bit too neat and tidy to be quite true. Nature never behaves in quite such an orderly, regular fashion. As one scientist observed: 'One can congratulate a player on having good luck, but if he is lucky tomorrow as well, and the following day, and the day after that, it is reasonable to become a little suspicious.'[3] It looks as if the eminent man of science – whether consciously or unconsciously – simply weeded out any peas that fell outside his framework and scrapped any results that didn't fit his theory. Then as now, scientists are prone to think, 'Something must have gone wrong, I must have been careless' if a result fails to match expectations. That's a cardinal error – but a very human one.

Biology is rather like a Russian nesting doll: each new discovery contains a new layer of complexity, exceptions and nuances. To study genetics in depth is to wander like Alice through Wonderland, exclaiming 'Curiouser and curiouser!' In fact, Alice and her journey down the rabbit hole into Wonderland are a very apt metaphor for genetics. The fictional Wonderland is a mix of the utterly absurd and what seems at first to be reasonable and familiar, but turns out to be bizarre: from the Cheshire Cat, who gradually fades away till only his grin remains, to the flamingos used as croquet mallets. Similarly, genetics may at first seem reasonable and clear-cut, like Mendel's peas, but as soon as you take a closer look, layer upon layer of complexity unfolds. Just like Alice's adventures in Wonderland, this too is an exciting journey of discovery that's far more enjoyable if you accept straight away that it may take slightly unexpected twists and turns.

All this doesn't mean that Mendel was wrong in expecting his peas to be tidy and organized. His theory is essentially correct where many traits are concerned, but those that depart from the general rule are both important and fascinating – and they play a vital role in helping us to grasp the problems and the potential of gene technology. To understand them, we'll need to take a more in-depth look at what exactly genes are.

About a century of scientific work and experiments, in many small steps, passed between Mendel's discovery of 'factors' in peas and the time when three scientists in Britain – Rosalind Franklin, James D. Watson and Francis Crick – succeeded in establishing what the DNA molecule looks like and how it functions.[4] It was these three who identified the double helix you see in school biology textbooks – the image of the DNA molecule that crops up everywhere an image of genetic material is required, in everything from Hollywood films to school science posters.

* * *

You may well wonder why every single one of our cells takes the trouble to invest energy in building and constantly maintaining extremely thin threads measuring a total of two metres – in every cell. If all the DNA in your body were joined together, the resulting thread would be four times the distance between the sun and Pluto.[5]

The thin thread is the DNA molecule, the double helix. The two metres are divided up into 46 pieces. Each of these is a chromosome, and together they form 23 pairs. You have inherited a set of chromosomes from each parent, one in every pair. The two chromosomes in each pair are alike, containing the same genes, though they can contain different variants. For instance, one may contain a variant linked with blue eyes,

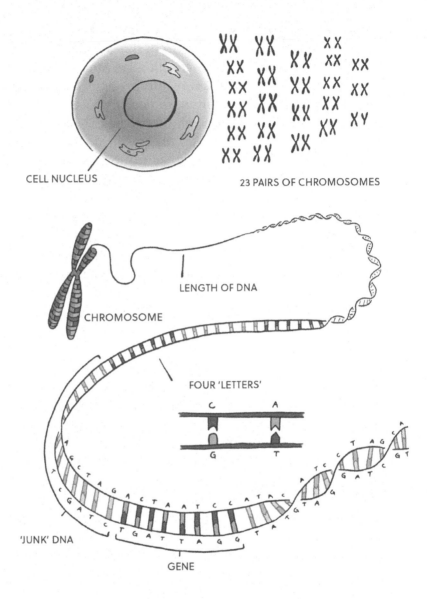

CELL NUCLEUS

23 PAIRS OF CHROMOSOMES

CHROMOSOME

LENGTH OF DNA

FOUR 'LETTERS'

'JUNK' DNA

GENE

Each cell contains a nucleus that in turn contains the genome. We have 23 pairs of chromosomes, each pair including one from each of our parents. A chromosome is a long DNA molecule, a ladder-like structure twisted into a spiral or helix. This double helix is made up of four substances: adenine, thymine, cytosine and guanine. The genetic code is thus made up of the letters A, T, C and G, each letter being located opposite another letter. Our longest chromosome is 249 million letters long, while the shortest has 47 million letters.

while the other contains a variant linked with brown eyes. The only chromosomes that deviate from this system are the two sex chromosomes. If you're genetically female, the pattern remains the same: you have two chromosomes that are alike, one from each parent – X chromosomes. If you are genetically male, however, one of these chromosomes is replaced by a smaller, shorter chromosome, the Y chromosome, which you've inherited from your father.[6] (Note 6 takes a more in-depth look at biological sex, which is more complex than just a matter of chromosomes.)

In principle, then, you have two sets of all chromosomes, and therefore of all your genetic material. There are exceptions: people with Down's syndrome have an extra copy of chromosome 21, either a whole chromosome or part of one. There are also people with more than two X chromosomes, and other variations exist.

* * *

Your genes are divided up among these chromosomes. And this is where we get to the core of the matter. Each gene does one thing: it gives the cell instructions on how to produce a single protein. There are layers of complexity here too, of course, several variants of what the genes do and the various stages on the way to a complete molecule. But the basic idea that one gene corresponds to one protein is a good start. You can compare the set-up to a library. Each cell contains the same collection of books. There are 46 bookshelves, each corresponding to one chromosome. Each shelf is full of books – genes, that is. The genes contain the genetic code, a length of DNA composed of letters, just as books contain words.

Once scientists had managed to work out the key (one gene corresponds to one protein), efforts to establish exactly what genes we have began in earnest. 1990 saw the start of the Human Genome Project, the aim of which was to map human genetic material as a whole.[7] If we could only read and understand the genes in our chromosomes, if we had access to the body's instructions, we would also understand everything that goes on inside us. It would be like having a combined recipe book and computer code, or so it was hoped. But the beginning of the Human Genome Project coincided with another major scientific breakthrough.

* * *

The exhibits of the National Museum of Scotland in Edinburgh include a stuffed sheep. At first sight, that's hardly surprising, as stuffed animals have featured in museums since the first natural-history collections emerged in the mid sixteenth century. And the odds on finding a sheep in a museum in Scotland, a country so rich in green hill pastures, haggis and woollen kilts, are scarcely high. But this particular sheep is unique. The birth of this Finn-Dorset ewe, on 5 July 1996, startled the whole world. For this is Dolly, the first mammal ever cloned.[8]

Cloning and genetic modification aren't the same. Cloning a living being means making a genetically identical copy of it and letting the copy develop into a new individual. Cloning plants is quite straightforward: you do that every time you take cuttings or plant a potato. But the process is more difficult in animals, and before Dolly many of the world's scientists were convinced that cloning mammals was out of the question.

To clone an animal, you remove a cell from its body and transfer the cell nucleus with all its chromosomes to an egg whose cell nucleus has been removed. When a foetus develops, all its nuclear DNA is thus identical to that of the original animal.

Now, over 20 years later, cloning is a commonplace though complicated process that goes on all over the world. If your dog or cat has died, you can order a clone of your pet based on cells from its body. The possibility of cloning animals has brought new hope for many of the world's threatened species. Moreover, combining gene technology and cloning makes it much easier to create bespoke identical laboratory animals – using the modified monkeys in Yunnan, for example. There is a consensus among the world's scientists that we shouldn't clone humans, and that hasn't happened as far as we know. But 2018 saw the first ever birth of a pair of cloned monkeys, Zhong Zhong and Hua Hua, two very appealing baby macaques who appeared before a world audience playing with plastic toys in their cage in Shanghai.[9]

* * *

The process of mapping the human genome took over a decade; the full results weren't published until 2003. But when, in 2000, the then US president Bill Clinton presented a first draft, his pride was evident: 'Today we are learning the language in which God created life [...] With this profound new knowledge, humanity is on the verge of gaining immense new power to heal.'[10]

Our genetic material, the DNA molecule, works like a book of instructions for our cells. Everything a cell does, and thus everything a body does, derives from the DNA molecule. A cell without DNA is like a body without a brain.

The long double helix is composed of four substances: adenine, guanine, cytosine and thymine (A, G, C and T for short). They are located opposite one another in a long chain, forming a genetic code that can be read by scientists, such as GTTTAGGACCCTGATCGA etc. (our longest chromosome comprises some 249 million 'letters'). It's quite easy to identify the genes among the multitude, as they always have the same structure: it's rather like reading a book, when you recognize passages of dialogue because they're in quotation marks.

Each cell contains a remarkable machinery whose sole purpose is to ensure that the genetic material, the 'instruction book', remains intact. This machinery comprises a large number of systems that serve to roll up the genetic material like thread on a spool when it is in use. These systems are constantly patrolling the strands of DNA, mending any gaps and ensuring that the enormously long chromosomes are always protected within the cells. Each time a cell divides, its genetic material has to be copied as well, and, what's more, divided up equally between the two new cells so that neither ends up with three copies of chromosome 14 but just one of chromosome 3, for instance. This is a well-orchestrated process, each step in which has been refined throughout the evolution of life to make it as smooth and problem-free as possible. But things do go awry even so. When the cell is copying its genetic material, individual letters can be omitted, swapped round or replaced by others, and small chunks can sometimes be copied several times in a row. These are mutations, tiny errors that are perpetuated at each further cell division. If a mutation occurs when an egg, sperm or embryo is forming, it can affect the future child's whole body.

Luckily, most mutations aren't harmful. The genetic code has inbuilt security mechanisms that make sure it can be read in most cases, even if a letter ends up in the wrong place – in the same way you can read 'teh' and 'pehraps' even when they're misspelt. A small number of mutations, however, end up in part of the genetic material where they do have an impact, generally a negative one. They can cause a cell to die suddenly or, in other cases, to start developing into a malignant tumour. In most cases the body can deal with such damaged cells, but sometimes they can cause major problems. Exceptionally, a mutation can alter the genetic code for the better. These lucky glitches are rare, but in the long term they shape us and all other species.

One example is the person living about 10,000 years ago in what is now Turkey who carried a mutation that enabled him or her to drink milk, even in adulthood, without getting a stomach ache. This mutation has since become common throughout Europe and the Middle East.[11] Gene technology gives us hope of being able both to correct some of the mutations that cause problems and to create new ones conducive to positive traits.

* * *

From Gregor Mendel's peas, via the structure of the DNA double helix and the mapping of the human genome, we have now arrived at a point where we can read the genetic code. We can see how genetic material from both our parents combines to give us the traits we bear. Now we have the capability to start mapping exactly which genes shape us. Moreover, the genetic code is universal, written with the same letters in all life on earth. Everything that differentiates us from bacteria, trees, fungi and chimpanzees is the result of individual, random mutations that have caused the branches of the tree of life to grow in different directions. But with CRISPR and other advances in gene technology, it's now possible for scientists to initiate changes themselves.

This is our first trip down into the labyrinthine rabbit warren of our genes. If you've been following the story closely, you'll realize that we haven't yet answered our secondary-school pupil's question about why all eyes aren't simply brown or blue. Before we can answer that, we need to take a look at how we can already save lives by combining CRISPR with our capacity to read the genetic alphabet.

7

Gene Therapy among Gilded Temples

Cheng Shufang is sitting up in a hospital bed with a drip-feed in her chest. Her short, perky black hair is sprinkled with grey. Though retired from her job in a government authority, she is tired out as a result of her illness. She speaks in a low voice, struggling visibly to sit up while I talk to her. At 53, Cheng Shufang has been battling an aggressive stomach cancer for the last eight months. So far it has resisted all attempts at treatment; in fact, she has metastases in her lungs and liver. Before her illness, she lived with her son in Canada, but on being told her diagnosis she returned to China for treatment here in Hangzhou.

The flourishing southern Chinese city of Hangzhou is a growing centre of science and technology. Despite its nearly 8 million inhabitants, it tends to be overshadowed by its more famous neighbour Shanghai, just over an hour away by bullet train. Yet Hangzhou attracts its share of tourists, being celebrated for its silk and tea and its beautiful natural surroundings, including the large West Lake and the forested hills that extend into the inner city. Gorgeous temples perched on mountaintops look down upon high-rise buildings. On the city's outskirts, next to forest, lies a small hospital: Hangzhou Cancer Hospital.[1] As the West Lake and the forest form a World Heritage Site, the hospital can't be extended any further, and no taller buildings can be built. This is something of a relief after a few weeks in a country whose endless ranks of skyscrapers and new-build complexes, crammed in cheek by jowl, have induced a feeling of dizziness that verges on vertigo.

When it became clear that Cheng Shufang wasn't responding to any other treatment, and that she had no other hope, she joined a small clinical trial here to test a new type of cancer therapy based on modifying her cells with CRISPR.

'I hope I'll recover. I hope God will bless me,' she says. She speaks a little English, but her doctor, Zhao Ruping, stands in as interpreter from time to time.

Zhao Ruping tells me what a poor condition she was in when she arrived at the hospital. The tumour had pressed her internal organs up into her ribcage, making it hard for her to breathe. It also caused fluid to accumulate in her abdomen, so that over a litre had to be removed daily. Dr Zhao says she had abandoned all hope; every night she thought she was about to die. All the patients taking part in the study are in the same situation.

'They don't have any other choice; their cancers haven't responded to any other treatment and they are in a late stage of the disease. So most of the time their status is not very good, and not all patients have responded well to the CRISPR treatment. Before treatment, these patients were nearly dead. After treatment they're alive – that's good,' he says. Exactly two weeks have passed since Cheng Shufang received her first infusion of genetically modified cells.

According to Dr Zhao, her overall condition seems to be improving and she's now breathing more easily. Preliminary tests show that the cancer has receded slightly and that her stomach tumour has stopped growing. It's too early to evaluate the result: the plan is for her to be treated with two more rounds of modified cells. This is a completely new, untried method, and Cheng Shufang is among the first people in the world to receive this type of treatment.[2] So I ask her whether she feels worried at all.

'I've seen some people worry, but I believe in the doctors. Before CRISPR I was very uncomfortable, so I'm not afraid of side effects. And these are cells from my own body, not someone else's,' she says.

Cheng Shufang's treatment is a variation on another type of cancer therapy that's also a recent innovation. The basic method, immunotherapy, is based on manipulating the body's own immune system. One of the functions of our immune system is to distinguish those cells in our body that are an intrinsic part of it – ordinary cells, to be left in peace – from alien cells like bacteria and viruses that need to be attacked. If our immune system attacks our own cells, the result is autoimmune diseases such as rheumatism or multiple sclerosis.

One element of the immune system is T cells, which search for chemical signals that tell them what to attack. Simply put, they sniff at cells within the body. There are substances that have a calming effect on the T cells, while others make them more active – like sniffer dogs that raise the alarm when they detect drugs, but ignore a perfume spray in your bag. Inhibiting mechanisms are important, as without them the immune system would overreact and damage the body, but they can also cause problems. Certain cancer cells are able to deceive the T cells by secreting substances that make the T cells ignore them.

In the 1990s, scientists discovered that they could shut down part of the T cells' inhibiting system, causing the immune system to start attacking these cancer cells as well. It took several years for American and Japanese scientists to develop these techniques, but a number of studies published in the early 2010s showed that this is an incredibly effective way to treat cancer in certain cases. This method, based on pharmaceutical drugs and chemistry, not gene technology, was rewarded in 2018 with the Nobel Prize in Medicine.[3] One of the laureates was accompanied to the Nobel festivities by one of the first patients to be treated, now fully restored to health. The method is a completely new way to tackle tumours, especially those resistant to all other therapies. It has given a big boost to doctors', scientists' and patients' hopes of a cure for cancer. However, just like other types of cancer treatment, immunotherapy can have side effects, particularly if T cells without an inhibiting system attack healthy cells in the body.[4]

When it became clear that cells could be easily and cheaply gene-edited with CRISPR, scientists began to think about how CRISPR could be used to make cancer patients' T cells even better at attacking tumours. At least ten studies in this field, using the CRISPR method, are under way in various parts of the world. And Cheng Shufang is among the first patients to trial this approach.[5]

The doctor leading the study, Wu Shixiu, meets me in a hospital conference room. It's not often I meet someone I'd call jovial – a cheery *bon vivant* – but 'jovial' is a perfect description for Wu Shixiu, cancer researcher. He has a ready smile, even when discussing serious scientific problems. Rather rotund and slightly short of breath, he jokes about both characteristics. He's delighted I'm so young, which he says he didn't expect. What a pleasure this is, how glad he is I'm visiting. Yes, he knows Lund, my home town. He's been there, he says: 'A famous university city, near the Bridge.'

We conduct our interview in hesitant English, and it's sometimes a little tricky for both of us to make ourselves understood. Nonetheless, Wu Shixiu constantly interjects little jokes and winks into the conversation. He seems thrilled that I'm here and sets aside plenty of time to talk to me, despite his assistant's discreet efforts to remind him that he has other tasks waiting for him.

The language barrier makes it hard to have a really nuanced discussion, and I'm not quite sure whether Wu Shixiu's evasive answers reflect language difficulties or the Chinese culture of conflict avoidance, which includes never really saying yes or no directly. After all, these experiments are controversial.

Wu Shixiu's original field of expertise is radiation oncology,

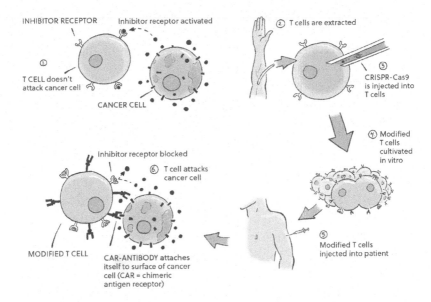

T cells are part of the body's immune system. Their targets include cancer cells. However, some cancer cells can release a substance that inhibits T cells. By removing T cells from a patient and editing them with CRISPR, scientists can 'switch off' the inhibiting substance. This enables the T cells to attack the cancer cells.

but at this hospital he's responsible for the full range of cancer therapies, which he tries to combine to obtain the best possible outcome. He focuses on solid tumours, rather than cancers of the blood or the like. Since April 2017, the clinic has also been trialling CRISPR immunotherapy on patients who have no other hope.

'CRISPR-Cas9 is a very good tool to reach our goals, and to edit immune cells. We feel we can give these patients hope – hope to prolong their lives,' he says.

He tells me that the method has been tested on cells under laboratory conditions and on animals, and that they ultimately decided to start trialling it on people. Over 30 patients have been treated so far. The first ones had cancer of the oesophagus: one of the ten most common types of cancer in China – 'if you're fat, like me,' says Wu Shixiu. Oesophageal cancer is also linked to alcohol abuse and smoking. This hospital, incidentally, is the first one I've ever visited where the smell of smoke lingers in the corridors and people in the waiting room sit with lighted cigarettes.

'CRISPR is simple and easy, and may be effective. And there's no severe outcome,' he says. The most common side effects they see are rashes and fever. Since the technique is still so new, nothing is yet known about possible long-term side effects.

The results of this research haven't yet been published, which means in practice that they're still hypothetical. Wu Shixiu says about 40 per cent of patients respond to treatment, but the effects are often quite short-term, extending lives by a few months. Sometimes the treatment also seems to make patients feel better in themselves, even if it has relatively little impact on their cancer. It's produced significant results in four people, and one of the first to be treated, one and a half years ago, is still alive and well at the time of the interview. Given how ill he was before starting treatment, that's surprising, says Wu Shixiu; in fact, it's almost miraculous.

Impressive though these results are – in patients who have exhausted all other options – the degree of success varies considerably. The approach is far from being the miracle cure many people hope CRISPR immunotherapy may offer in future. However, it's hardly surprising that the technique and its potential are still sketchy, since it's still at such an early stage of development, says Wu Shixiu. If they don't trial this approach, they can't learn more about it. At any rate, he thinks the results are promising and is now applying the method to patients with pancreatic, breast and bowel cancer, one of whom is Cheng Shufang.

The process begins with the doctors removing 100 millilitres of blood from the patient. It is sent by bullet train to a laboratory in Hefei, a few hours away, where the T cells are separated out and edited with CRISPR. The technique used is called 'gene knockout', meaning simply that CRISPR disables a gene governing one of the inhibitors. This means the edited T cells aren't held back by that inhibitor, so they attack the tumours. Knocking out a specific gene is something for which CRISPR is ideally suited.

The next stage is to let the modified T cells grow and divide until there are enough to make a difference. That generally takes about two weeks, or slightly longer if the patient is in a very poor condition. Once the cells are ready, they are sent back to the hospital by train, and the patient is placed on a drip containing a substantial dose of his or her modified cells. Since these cells have been edited outside the patient's body, there's no risk that any other cells or other body parts will be modified. That's one of the main advantages of this technique.

At the same time, in-vitro editing also poses problems: it means that this will always be a slow and expensive method. Devoting long laboratory hours to cells from each individual patient, followed by the weeks the cells need to grow, demands far more resources than conventional cancer therapy. However, scientists hope that in future doctors will be able to inject tumours directly with CRISPR, with instructions to the cancer cells to kill themselves.[6]

That would be a faster method than immunotherapy. After patients at Hangzhou Cancer Hospital have received their first round of treatment, they have to wait: first to see whether the treatment seems to be helping and whether it's worth going on to a second round, then for their third round of treatment – if they haven't died of cancer by that stage. Most do.

Since the method is new and untested, scientists are applying it only to patients for whom no other possible cure exists. It would be unethical to trial it on other patients, Wu Shixiu thinks, even though there are others who might be helped. Once there's a consensus on how to apply the therapy to a late-stage cancer patient and it's been shown to be effective, they can start thinking about using it at an earlier stage. At the moment, he says, that would be too controversial; besides, they can't yet say whether it really is effective.

The risks involved in a clinical trial like this one are of a different order from those posed by projects like He Jiankui's embryo experiments. What's at stake isn't the future of the human race, but the risks incurred by individual patients: the risk of suffering and unexpected serious side effects, the risk that the scientists may inadvertently speed up the growth of a cancer instead of slowing it down, and so on. But these risks need to be weighed against the fact that the patients have incurable tumours which have failed to respond to any other treatment. There's no hope of saving them in any other way. Nonetheless, studies like this are far more heavily regulated and monitored in Europe and the United States than in China.

The procedure Wu Shixiu had to undergo before starting the study was swift and streamlined. The ethics committee tasked with assessing his proposal took just one afternoon to approve the clinical trial. His work had to be approved by the hospital and the local authorities, but it wasn't submitted to the national authorities, as it would have been in the United States or Europe. 'The T cells are the patients' own cells, so we can use them more boldly, even when we change them,' he says.

Since the United States was in the vanguard of research in this field, at least a few years ago, one might reasonably have expected studies on humans to begin there. Yet China has very rapidly outpaced both the United States and Europe in nearly all areas of gene technology. American scientist Carl June, who leads one of very few similar studies in the United States, is on record as saying the Chinese system is like 'the Wild West', with a mix of very high-quality research and studies of a much lower calibre.

I ask Wu Shixiu if he thinks CRISPR immunotherapy will make a big difference to future cancer treatments.

'I don't know,' he says. 'Because of the debate in the West, it's not being used in clinics. If you don't use it in human beings, then you'll never know if it's effective, what the side effects are, how to improve it. So after accepting that we can edit cells from patients, that it's safe, it's reasonable, we can monitor what happens after the treatment and then form a common consensus. Then we can find more combinations with new technologies, things that we can use to improve it. Then we can see its potential. Now is just the beginning.'

He thinks I'll probably live to see wonderful results from CRISPR, but says that at just over 50, he's too old. During our conversation, he often touches on the differences between China and the West, so I ask him if he thinks Westerners are too cautious. With a laugh, he replies: 'I don't know. They see every part of the body as very mysterious, as...' The word he needs eludes him for a moment and he asks me if I'm a Christian. No, I reply. A Muslim, then? No, I'm an atheist. 'Ah, an atheist...' he says with a chuckle. He continues to search for words until I suggest 'holy'.

'Yes, exactly, holy. Like a king. They [Westerners] consider the body very, very holy, which makes them cautious. That may be good, but according to Eastern knowledge, it's not wise to view the body as an emperor. Every patient wishes to survive longer. They're afraid of immediate death. If they live for a longer period [...] it's OK.'

Wu Shixiu thinks people worry too much about the risks of gene technology. This new technology has been much debated by scientists in the United States and Europe, and many people, both the scientific community and the authorities, are cautious about starting trials on patients, even those who have no alternative. But Wu Shixiu rejects the idea that this might involve risks.

'This is CRISPR in somatic cells. It cannot be transferred to descendants. These cells have relatively short lives; they disappear very quickly.'

I ask him what he would like to say to those with concerns.

'Patients are scared of death, not scared of the treatments: that is important. People in good health are afraid of CRISPR, not of death, because they don't think they're going to die. The patient has the right to say: "We need this."'

At present, China is almost alone in offering this choice. Although other types of anti-cancer immunotherapy are used elsewhere, China is virtually the only country where scientists are experimenting with CRISPR, though some studies are now beginning in the United States as well. Although it's impossible to know exactly how many medical trials are under way worldwide, the American website Clinical Trials (clinicaltrials.gov) is a good starting point: it shows all the trials listed by US authorities. These aren't just studies conducted in the United States, as the database is now used throughout the world. Clinical Trials lists about 230,000 ongoing studies in 195 countries. A search reveals that about 20 CRISPR trials involving several types of cancer have got under way in China over the last few years. All of them are essentially variants of the study led by Wu Shixiu and involve changing the way that the immune system (particularly the T cells) reacts to cancer cells. The first study seems to have started in 2015, but most began in 2017 or 2018. So far, the results are limited. However, many other studies being conducted in China don't appear on the list.[7]

'There are many, many. To my knowledge, in our province there are more than 20 researchers doing this. Some by themselves, others cooperate with companies. [...] Sometimes they have venture capital to fund the research. [...] The number of clinical trials is increasing,' says Wu Shixiu. But his study is the only one in this region to be registered in the American database.

Gene technology is both new and complex, so I ask Wu Shixiu whether the patients really know what they are letting themselves in for. 'Yes, they do, absolutely,' he replies swiftly. 'They know, they come to seek it. Most of them are not in our hospital, they come from everywhere. China, but also from America, India, Poland and many, many places. Many people contact us, but they cannot always tolerate long-distance travel.'

Cheng Shufang's doctor, Zhao Ruping, is in the room during our conversation and joins in at this point. He's responsible for dealing with all international enquiries.

'Seven days ago I received a telephone call from Denmark,' he says. Recently they treated a patient from India with success. Taking out his mobile, he shows me all the emails he receives each month from people keen to try the treatment. He is contacted by patients, family members, doctors and hospital administrators. 'Sometimes I don't even have enough time to reply,' he says. These are desperate people with no further hope of being restored to health through conventional therapies.

The CRISPR-based immunotherapy itself is provided

free of charge to patients whose state of health allows them to make the long journey to this hospital. The doctors also need to approve them as suitable, since they are to take part in an ongoing study. But everything else has to be paid for: their hospital stay, medical care, drips, tests and so on. Zhao Ruping doesn't want to give exact prices, as they vary, nor does he really want to say what the maximum cost might be. However, the cheapest option, for patients who are well enough not to need a hospital stay and who need relatively little medical care apart from the CRISPR treatment itself, is 20,000 yuan (well over 3,000 US dollars in the summer of 2021). For those patients who need a significant amount of medical care – all of them in practice – the cost is far higher, says Dr Zhao.

'They put so much faith in us. They think we are gods,' says Wu Shixiu.

This illustrates the other ethical issue raised by experiments of this kind. Since Wu Shixiu and the other doctors and scientists haven't yet published their results, nobody knows how effective their treatment really is. Everything I observe and find out indicates that things are as they seem, that the team provide the care they say they provide, and that they can help some of their patients. However, in recent years many 'clinics' have sprung up worldwide, offering 'experimental methods' that are actually nothing but quackery. Patients are injected at best with innocuous saline solution, in the worst cases with toxic cocktails of cells and chemicals that have no effect.[8]

It's hard to gainsay doctors who want to give patients hope of a cure. The Hangzhou Cancer Hospital doesn't charge exorbitant fees, and the results Wu Shixiu presents to me – fewer than half of all patients are helped by the treatment – suggest that they are being honest. But even if patients are injected with flawlessly gene-edited cells, we still don't know for sure whether this method works. There's a risk of investing time, money and hope in a mirage. Conversely, any remedy or relief is better than none, and without hope and clinical trials like this one, it would be impossible for science to progress.

This path inevitably involves suffering and crushed dreams for terminally ill patients trying new treatments, even if the ultimate outcome is a great success.

There is a hope, a chance, that gene-editing immune cells with CRISPR may provide a new, effective treatment for cancer. That would be wonderful. The technique could also prove to be another weapon in the arsenal that includes radiotherapy, chemotherapy and other forms of immunotherapy already available to physicians. There is also another type of genetically modified immune cells used to treat cancer, known as CAR-T cells. The most likely scenario for CRISPR is that rather than offering a new panacea, it will become another tool in our toolkit.

'I think CRISPR-Cas9 makes editing of genes more precise and more convenient. [...] But maybe in the future there will be new techniques to edit genes. [...] CRISPR-Cas9 is just one stage of the science,' says Wu Shixiu.

* * *

I return home from Hangzhou Cancer Hospital with a fairly positive impression of the trial and its potential, even though it's clearly problematic for studies like this to be conducted on the basis of snap decisions, and with so little ethical oversight. But when I contact Wu Shixiu in the autumn of 2019 to ask some further questions, I hear that the study has been stopped. No more patients are being treated. This is because of the incident with the babies, Wu Shixiu writes, referring to He Jiankui's experiment. The situation in China changed fast after the gene-edited twins were born and the case achieved notoriety: although Wu Shixiu's study had nothing to do with editing the genes of human embryos, most projects involving the acronym 'CRISPR' and human beings seem to have been shelved. Cheng Shufang has died, despite the promising results of the treatment, but I'm told that two other patients have made a full recovery. Yet the team hasn't yet found a scientific journal that's prepared to publish their results.

So far, Hangzhou Cancer Hospital's experience is just that: experience and inconclusive evidence. But the fact that CRISPR is effective, cheap and easy to use means that its potential goes far beyond cancer therapy – temporary setbacks notwithstanding.

8

Panacea

'Absolutely everything.'

That was scientist Irina Conboy's reply when *New Scientist* asked her what diseases might be treated with the help of CRISPR. Conboy's own research at the University of California, Berkeley, focuses on conditions like Parkinson's disease and muscular dystrophy, but many scientists working in other fields would give equally enthusiastic replies.[1]

Think of a serious disease of any kind. More than likely, there's a scientist somewhere in the world pondering right now how CRISPR, or another gene technology, could be used to cure, alleviate or shed light on that very disease. And this isn't just about gene-editing our own cells. It's also about using genetic technology to produce laboratory animals that are better adapted to scientists' needs, to understand the workings of viruses and bacteria, or to invent new medicines or new ways to diagnose diseases. Or simply to produce cheaper medicines using gene-edited bacteria. This technology is also being used at least as much now in fundamental research into the workings of our own bodies and how different genes affect one another. A sense of vast potential currently pervades much of the medical research community – as if scientists were artists suddenly presented with a brand-new colour in their palette.[2]

'Potential' is precisely the word you need to bear in mind when you hear about their dreams: all very promising and enticing, but nothing's yet certain. That also means the scientific community doesn't yet know where the boundaries lie, or which of the applications scientists now dream of will be far harder to achieve than they think – maybe even impossible. But then we don't know which applications will turn out to be unexpectedly straightforward either.

You know the feeling of being newly in love. Not just the butterflies in your stomach, but those shimmering castles in the air: this relationship might just be the best one ever; it might just give you everything you've always wanted, from the perfect apartment – always kept tidy, without any squabbles – to holidays full of passion and intimacy, during which the two of you leave the rest of the world behind you. All that potential is there at the start of every relationship, but it's never quite the same in practice. Even the most wonderful relationships differ from a newly smitten lover's daydreams, and none are without friction. So far, the scientific community is still in the early stages of its romance with CRISPR and views it through rose-tinted glasses. That's bound to change.

There are over 6,000 diseases caused by damage to a single gene, some of which occur spontaneously, while others are passed on by parents to their children. They include cystic fibrosis, Huntington's disease and haemophilia.[3]

One of the most heart-warming and hopeful examples of the positive impact that CRISPR can have comes from a children's hospital in Memphis, Tennessee. When scientists there removed stem cells from the bone marrow of eight newborn babies, the odds were against them. The babies had a serious genetic disorder, a form of severe combined immunodeficiency called SCID-X1, but often referred to as 'bubble-boy disease'. Children with this disease have a severely compromised immune system, so a common infection can make them seriously ill and even kill them. They have to be isolated to avoid infectious illnesses; in the past, they were often protected by a plastic bubble around their head, to ensure that they only breathed filtered air.

Children with SCID-X1 can sometimes be helped by a transplant of bone marrow (from a sibling or parent) containing healthy stem cells that can kick-start their immune system, but that doesn't work for everyone. Given the very serious nature of the disease, many attempts have been made to develop drugs and treatments, but none of the results have proved very effective. While some children have recovered, others have developed leukaemia.

This time everything seems to have gone well. The eight babies' bone marrow was modified using CRISPR, which was delivered to their cells using a new system. Just over a year after treatment, all eight were able to go home to their families, and their immune systems now seem to be working normally, without any serious side effects.

'They are all toddlers now, exploring life and attending daycares,' said Ewelina Mamcarz, the doctor leading the research project, in an interview with *Nature*.[4]

* * *

Using CRISPR to cure 'absolutely everything' faces several huge challenges. One is the gene-editing process itself. The first stage – removing cells from the human body to fight cancer or treat children with damaged immune systems, for instance – is relatively straightforward. It's easy to check if the genes have been correctly edited, and easy to avoid affecting other cells in the body. But it takes a long time to culture enough gene-edited cells to make a difference once they've been returned to the patient's body. Moreover, this method only works for certain diseases. You can't remove a person's eye to modify it, and then pop it back into its socket.

The other method is to let the gene-editing process take place inside the patient's body. This involves using some form of transport to deliver the CRISPR system to the body. The most usual way is to use harmless viruses as vectors. This is the method that made Jesse Gelsinger's immune system overreact, but scientists have since got much better at finding safe techniques.[5] The challenge now is to modify the right cells in the patient's body, and to make sure that enough of them have been altered. Diseases differ significantly in this respect. If you only need to edit a few cells in one organ, that's much easier than editing all the cells in a muscle, for instance – let alone all the cells in the body.

In November 2017, a patient in California became the first ever person to be injected directly with a gene-editing tool to modify genes in his liver.[6] He had a rare genetic disorder called Hunter syndrome. People with this disease cannot break down certain harmful substances in the liver, so these accumulate in the body, causing fatal poisoning in the long run. Sufferers have just one flawed gene. The purpose of the trial was to insert a new, healthy variant of that gene in the patient's liver cells.

Another aspect of the treatment was that it was designed to splice in the new gene in a location in the patient's genome where it was almost bound to be activated, to make sure that it would actually benefit him. There's always a risk with gene technology of inserting or editing genes which then have no effect, because other parts of the patient's genome control which genes are active in each cell. It's as if a gene were inserted into a library book that was never opened again.

A single injection isn't enough to modify a significant number of cells. However, since the toxic substance is only present in small quantities, it's enough to have a few liver cells able to process it. Follow-up a year later indicates that the patient's liver is now starting to break down the toxins to some extent, and that he's generally in good health. The few other patients on whom the treatment has been trialled are also showing promising results, according to the firm developing the method. Another firm has recently started injecting CRISPR into the eyes of children and adults born with a genetic eye disorder. It's hoped that the gene scissors will correct the mutation in enough cells for the participants in the trial to recover their eyesight to some extent.[7]

In the case of the disorders discussed so far, starting treatment later on in life, or at least after birth, works well. But what can be done to tackle diseases that begin affecting the foetus during gestation? With some of these, the solution will involve applying gene technology to both the foetus and the mother-to-be during her pregnancy. British scientists are currently looking at gene-editing foetuses with certain disorders in the womb, if amniocentesis has revealed genetic anomalies. Such intervention would take place early enough to counter problems in foetal development, but not so early as to affect the individual's future eggs or sperm. That means that any changes wouldn't be passed on from generation to generation, unlike those made by He Jiankui.[8]

The cases in which CRISPR will achieve results fastest will presumably be those involving known genetic disorders that can be treated by editing or replacing a single mutant gene. However, many scientists are also looking at how gene technology can be applied to more complex diseases, disorders that aren't caused by a single defective gene, or that attack us from the outside.

I was born in the mid 1980s, late enough to have missed the worst of the HIV and AIDS crisis, but early enough to have seen the huge impact of antiretroviral drugs in treating patients. The disease aroused so much fear that much of my sex education at school focused on protecting oneself against HIV.

That fear has receded to some extent, and fortunately some of the stigma surrounding HIV has gone. Yet nearly 2 million people are still infected every year, and about 38 million people are living with the disease.[9] Although antiretroviral drugs are very effective, they are also costly, and they place heavy demands on both the individual and society as a whole.

The HIV virus is more than usually stealthy. It resembles other viruses in the way it works: by attacking a cell. The cell is obliged to produce new viruses so that more cells can be infected. However, the HIV virus doesn't infect just any cells, but specifically those in the immune system. As a result, the body's own system of defence is unable to identify and combat the virus, as it does in the case of viruses like the one causing the common cold. Moreover, the virus gradually breaks down the individual's immune system, leaving them very vulnerable to opportunistic diseases.

A number of teams of scientists are seeking ways to use gene technology to actually cure people of HIV, rather than merely controlling it, as antiretrovirals do. Japanese scientists seem to have found a way to prevent infected cells from producing new viruses. That would mean that the patient could be fully restored to health, as the infected cells in their body die and are replaced by new ones. The method seems to work well on human cells in vitro, and the very first trials on living people have now started, but it will take a while until these methods can start to be used on a large scale. Other scientists are trialling a range of variants based on the same idea, and there have been rapid developments. It looks very much as if gene technology – whether CRISPR or some other technique – will come up with an effective cure for HIV and AIDS.[10]

The development of a cure for AIDS also sheds light on certain problems with this technology. Many of the people living with the disease today aren't taking antiretrovirals, because they simply can't afford them or they live in countries where such treatment isn't available. A genetic cure would presumably be too expensive for many of the people who could benefit from it, even though such treatment would be much cheaper than antiretrovirals in the long term.

Who should get access to new technology, who pays for it – and who decides which technology is to be developed, and which diseases scientists should seek cures for?

* * *

One disease that illustrates these issues particularly well is sickle-cell anaemia, a genetic disorder that kills over 100,000 people every year, while 4.5 million people live with the illness. Despite these figures, it has a relatively low profile, particularly in Europe, because most of the people it affects live in central or West Africa, or have their genetic roots there.

In genetic terms, sickle-cell anaemia is an intriguing device used by the human body to protect itself against another disease – malaria. When a person is bitten by a malarial mosquito, the single-celled organism that causes the disease enters that person's circulation. Once there, it infects the liver and the blood cells, and it can then infect the next mosquito that bites the infected person. Malaria has plagued humans for about 50,000 years, causing millions of deaths every year.[11] So it's not surprising that humans have been subject to powerful evolutionary pressures to develop some form of anti-malarial defence.

A few times in the course of history, variants of a particular mutation that protects the carrier against malaria have made their appearance.[12] The mutation is a change in the gene that alters the haemoglobin in the red blood cells, making it far more difficult for an infection to spread through the body. The downside, however, is that the red blood cells change shape from Frisbee-like discs to slightly shrunken crescents. They are rather like the old-fashioned sickle on the Soviet flag, hence the name 'sickle-cell anaemia'. As a result, the blood cells have less capacity to transport oxygen in the bloodstream and are more prone to clumping together and causing blood clots. These clots cause agonizing pain. The fact that a mutation as harmful as this has spread and become common nonetheless is eloquent proof of just how serious a disease malaria is.

The vast majority of people bearing the mutation don't have major problems. This is because they've inherited a normal gene from one parent and a mutant gene from the other parent, meaning that their blood cells aren't seriously affected. They are better protected against malaria than most of the population, while the normal gene compensates for many potential problems. Genetically speaking, they have the best of both worlds. As a result, a third of the population in certain parts of the world carry a mutant gene.

Problems arise if two people who both carry the gene have children together. About a quarter of their offspring will have two mutant genes and no normal genes. (Now we're back to Gregor Mendel and his peas, and the reason that some people have blue eyes and others brown.) These children will have sickle-cell anaemia. The problems mostly start when babies reach five or six months of age, though sometimes the disease appears later in life. Those affected have anaemia, and they can suffer from swollen hands and feet, blood clots that inflict excruciating pain, and a battery of other symptoms. Some are particularly badly affected, while others have milder variants of the disorder. Even if sufferers get medical care, their life expectancy lies between 40 and 60, but a tenth of sufferers die before reaching the age of 20. The odds are much worse in countries where healthcare is less well developed.

Sickle-cell disease is relatively common in the United States, as many descendants of the slaves abducted and trafficked to America carry the mutation. Despite its widespread incidence, the fact that nearly all of those with this disorder are African Americans means that relatively little effort has been put into developing a cure – until now. It now looks as if gene technology may offer an effective way to put paid to the disease: experiments with gene therapy on mice have had very good outcomes. Early clinical trials on a few patients have also produced promising results. There are several research teams working to develop a cure with the help of CRISPR, and everything suggests that they will be successful.[13] Yet problems arise at the interface between medical science and real life.

* * *

Michael Friend was working in real estate when his wife suddenly fell seriously ill. She had sickle-cell anaemia but hadn't had any symptoms before, and neither she nor Michael knew a lot about the disease. Now their lives were turned upside down.

'What I found out very quickly is that treatment for adults is horrible. There were issues with getting proper care, getting her the level of pain medicine she needed,' he says.

Many patients testify that it is hard to get treatment, that they aren't taken seriously by doctors and nurses, and that it is difficult to get effective pain medicine. According to a number of studies, the main reason for this is that African American patients are just not taken as seriously as white people. They have to wait longer for treatment and are treated worse – and on top of that they are often prescribed lower dosages of painkillers for the same ailments.[14]

Michael Friend decided to leave his job, support his wife and advocate for sickle-cell patients. His first step, apart from taking care of his family, was to contact other sufferers. They sent a collective letter to the then US president, Barack Obama, urging him to focus on the need for new drugs. They met Obama in 2011, and ever since Michael Friend has run the Minority Coalition for Precision Medicine. One of the main aspects of its work is bringing together those keen to develop remedies with pastors and other community leaders, to find volunteers for clinical trials. What they want to do is match 100,000 scientists with 100,000 churches, and to do this worldwide, he says.

Many African Americans have a sceptical attitude to medical care in the United States and are reluctant to volunteer for clinical trials. There are many reasons for this. Above all, American history has numerous examples of African Americans being exploited for medical research, infected with diseases or having medical care withheld altogether. The most notorious example is the Tuskegee study involving hundreds of Black men with syphilis, which ran from the 1930s to the 1970s. The participants believed they were being treated, when in actual fact they received no treatment at all, even after effective drugs had become available. What the scientists were actually interested in was what happened when the disease was left to run its course untreated.[15]

Another well-known example is that of the HeLa cell line. Henrietta Lacks was an African American who died of cervical cancer in 1951. However, a cell sample which scientists took from her tumour remains alive and is still being used for scientific purposes. These cells have contributed to some major scientific discoveries, but Henrietta Lacks's family remained unaware that they were being used for a long time.

Nor did they derive any financial benefit from the many drugs developed using her cells.[16]

So a lot of Michael Friend's work involves bridging the gap between the scientific world and the African American community, but he insists that the exchange has to be reciprocal. There must be African American scientists in the teams he works with, and if a discovery is patented, African American interest groups must have a share in them. He believes that a cure developed under these conditions could help heal relations between communities. African Americans will feel that someone is finally listening to them and focusing on their needs, without trying to exploit them. 'That will be a game changer,' he says.

* * *

In Ancient Greek mythology, Panacea (Panakeia) was a goddess representing perfect health. Her name later came to stand for a mysterious substance with the power to cure all ills. The same alchemists who sought to make the philosophers' stone also strove to concoct a potion that would provide a universal cure.[17] In fact, we are still hoping for a wonder drug to cure all our disorders. Sadly, that's bound to remain a dream. But CRISPR and other gene technologies will undoubtedly revolutionize medicine, restoring health and bringing relief to thousands of people, a prospect which has prompted the world's scientists to dream ambitious dreams that may be over-optimistic.

Some of those dreams will be fulfilled, others will prove unattainable, and for many – in fact, presumably all of them – the journey from bench to bedside will take far longer than those suffering today would like. I am optimistic about the potential of research in this field, but it won't cater for all our needs. That isn't really a problem. Antibiotics didn't offer a panacea either, but I certainly wouldn't want to live in a world without them. It's possible to be optimistic without ignoring the possible pitfalls. Each time this new technology takes another step forward, the chances increase of saving someone's life, of making life better for a family somewhere.

9

Avatars

It's easy to get bogged down in theoretical arguments and hypothetical cases when we talk about how genetic technology could change the world. But both the opportunities and the major ethical challenges it entails are brought into clear focus by a boy called Mason and his family.

When Mason reached the age of three he started to have tantrums. The once happy little boy would suddenly go berserk, screaming, lashing out and tearing around. Each outburst would last for half an hour, sometimes longer. Mason's father, a tall, strong man, would have to hold him in a bear hug to stop him hurting himself or others. He would hold Mason and talk to him gently until the child calmed down. 'Then, suddenly, it'd be over. And he'd look up at me and he'd be like: "Daddy, I'm a good boy." You know, you could see he felt guilty about it. But it just happened. It was traumatizing,' says Mason's father Chuck.

Like all parents, Chuck and his wife Malia tried to find out what was wrong. Was it their child's diet? Eating certain kinds of sweet treats gave Mason a rash and made his temper even worse. They noticed other problems, too: Mason's speech development was slowing down and he would sometimes deliberately bang his head. In the same year, Mason had surgery that involved inserting a tube into one ear: he'd suffered from one ear infection after the other. At that point, the doctors discovered a large tumour on the left side of his face. It had affected his ear canal and expanded into one of his sinuses. It was a plexiform neurofibroma, an inoperable benign tumour growing on a nerve. Was this the root of all his problems?

The tumour came as a shock to the family, though it wasn't entirely unexpected. Such growths are a common symptom of the genetic disease Mason suffers from, neurofibromatosis type 1 (mostly known as NF1).[1] Mason was diagnosed when he was a year old, after his parents had noticed small patches of pigment on his bottom and legs. These patches looked quite cute, but as more and more of them appeared, Chuck got worried and started searching for information online. The *café au lait* spots turned out to be an early symptom of a disease he'd never even heard of.

NF1 isn't a particularly rare condition, though it's not widely known. About one in every 3,000 children is born with the disease. About half of these have inherited a flawed gene from one parent, just as with many other genetic diseases. In the other half, the mutation has occurred spontaneously: neither parent carries the disease. That was what had happened to Mason.

NF1 arises from an anomaly somewhere in the neurofibromin gene, an unusually large gene located on chromosome 17. The gene is found in all mammals and appears to play a major role in foetal development. The world's scientists don't yet know exactly what it does, although it's active both at the foetal stage and in adults. It seems to be involved in neural development; it fights tumours and appears to affect hundreds of other bodily processes. NF1 is vulnerable to many types of anomaly because it's so large, over ten times the size of the gene for blue eyes. Individual letters in the genetic code can occur in the wrong place, with an A being replaced by G, or a T by C, for instance. Alternatively, sizeable sections of the code may disintegrate or be missing altogether. As the gene plays a role in so many different processes, each mutation has unique consequences.

Scientists estimate that there are some 4,000 variants of NF1, each with slightly different symptoms.[2] Some people hardly notice they have the disorder, while in others it affects every aspect of life. The disease can result in neural damage and problems with the legs. Babies and adolescents are at risk of sudden death. Many sufferers develop tumours and large growths on their faces and bodies. Because of that, many of them don't want to be seen in public and live secluded lives. That's one of the reasons the condition remains little known.

Moreover, the same variant, the same mutation, can produce different symptoms in different individuals. The reason for this is presumably that the defective gene interacts both with the other genes in the individual's genome and with the environment. Chuck tells me about families in which all the children have inherited the same mutation from one of the parents but show quite different symptoms. There is a tangled web of symptoms which scientists and physicians are trying to identify and treat. It's almost impossible to get an overall picture of the disease – in contrast to Mendel's well-behaved peas.

For Mason, the disease has meant an autism diagnosis and a delay in his schooling. He's also at great risk of developing more tumours in future. When I meet Chuck, Mason is a mostly cheerful and lively eight-year-old. His facial tumour is the size of a lime but has stopped growing. He's now having slightly fewer tantrums, and after each incident he apologizes to his teacher, schoolmates and parents.

'You can't even imagine how frightened and upset he must feel when he's had one of these episodes,' says Chuck. 'I mean, he'll completely demolish a classroom and then afterwards [...] he knows he's done it and he can't stop. You know, you can't discipline him because it's not his fault. It's the chemistry that's doing this to him. Can you imagine?'

Mason isn't the only one for whom the disease has been a burden. As we talk, Chuck comes back again and again to how exhausting it has been for the whole family. Malia took a year off from her job to focus on helping Mason catch up at school and to bring the family together again. Now Mason is only a year behind his schoolmates, and he's up to date with his year group in maths and science, the subjects Chuck is most interested in. Mason's little sister Alexandra also faces problems. She was born with only two fingers on one hand and had kidney complications. There were times when both children were in hospital concurrently. But now they're doing better, and Malia is back at work again. Chuck jokes that they'll probably be able to redecorate their home and put in new skirting boards in a few years' time, but until then the results of Mason's outbreaks will have to stay where they are.

Chuck jokes and laughs a lot. He says humour is the only way to cope with it all.

'I'm not gonna lie, I cry a lot, typically in the car ride home. It's just, you can't do it in front of family, you have to keep strong in front of them, and sometimes it's all too much.'

* * *

The reason I meet Chuck isn't first and foremost to talk about Mason and the heart-rending challenges facing the family. I'm here to talk about pigs. About what happens when CRISPR meets real life.[3]

Chuck (Charles Konsitzke, to give him his full name) is the associate director of the University of Wisconsin-Madison's Biotechnology Center, a few hours' bus ride north of Chicago. After Mason was diagnosed, Chuck started reading up about his condition and tried to identify the scientific bottlenecks that had been holding back the development of better treatments – and maybe even a cure. At a meeting, he ran into Dhanu Shanmuganayagam, a scientist working at the same university who had been researching cardiovascular diseases for many years and happened to be looking for an interesting new research field to get his teeth into. Together, they decided to use CRISPR to find a totally new solution to NF1.

I accompany them to a research centre situated in the fertile clay fields just outside Madison. We arrive at a place that looks like an ordinary farm, but you have to go through a complicated procedure before you can enter. I go into a changing room and remove every item of clothing I have on. Then I have to shower thoroughly, lathering myself and washing my hair. Leaving the shower, I enter another room where I find dry towels and plastic boxes containing clothes in different sizes. I'm not even allowed to bring my hairband in with me. The only item I can bring in is my glasses – after cleaning them meticulously. I find a new hairband to put my hair up, and rummage around in the boxes of clean clothes. I dig out some underwear and a lime-green jumper, then find some workman's trousers that are far too long for me, and stuff the trouser legs into tall green gumboots. When everything else is ready, I put on some blue plastic gloves. In the meantime, my notebook and voice-recorder have passed through an airlock. I'm not allowed to bring my camera in.

This whole procedure has been put in place to protect the centre's pigs against bacteria and viruses, which we humans might otherwise bring in with us from the outside. For the pigs here aren't like other pigs. The first one I meet is Tank, a massive boar who cheerfully comes forward to say hello when we reach his sty. I get to pat his back, and he nudges my plastic gloves with his snout. Though curious, he's a little shy and seems more interested in playing with the plastic balls on the floor than in getting to know me.

Just like Mason, Tank has NF1. He has a specially designed mutation on the NF gene, a gene modification which scientists gave him when he was nothing but a single-celled embryo. Just like millions of other lab animals worldwide. At his birth in 2016, he and his three sisters were the first pigs in the project to carry the disorder. Tank also became the first boar with NF1 who was trained to donate sperm to future generations of NF1 pigs. I discover that the pig-breeding world has a strong focus on sperm. The sows are never given names, which is why I don't find out what Tank's sisters are called. The boars aren't named either – until they've been trained to mount a dummy sow made of steel and pump out sperm to be collected by waiting scientists. The scientist who is first to 'milk' a boar successfully gets to name him. In Tank's case, however, there was universal agreement that that honour fell to Mason. The scientists joke that Tank will be part of an armoured division in the war on NF1.[4]

Tank may be the first, but Dhanu and Chuck plan for many more pigs with NF1 to be born here. For pigs may be the key to finding a cure. Although the neurofibromin gene exists in all mammals, pigs are more appropriate as test subjects than rats or mice. They're a similar size to humans and live for a long time. They're also clever enough for scientists to be able to test how the disease affects their intelligence and behaviour.

In developing new laboratory animals, scientists normally aim to identify a standard variant of the disease being researched, but that wouldn't help in the case of NF1, as it takes such an incredible variety of forms. A drug that cured one patient's symptoms might be useless to another. Tank has quite a mild form of NF1. The scientists wanted to be sure that any flaws were the result of their methods, not of the disease itself. Now they've developed the process more fully, they can begin to develop pigs with more serious symptoms and make a start on the real project – creating pigs that mirror children with the disease. Rather than develop one or a few different pigs with a general form of the disorder, the scientists plan to develop piglets with exact copies of the defective NF gene from different children.

They refer to the pigs they mean to develop as 'avatars', a Sanskrit word referring to the physical manifestation of a god, the form that a deity assumes when on earth. Having been used as an IT term for the tiny image representing the user of a program or game, and then associated with *Avatar*, the hit film that came out in 2009, the word has increasingly come to mean 'representative'. That's how the scientists view the pigs in their project. Modified with CRISPR, these animals become representatives of sick children, patients. This enables the scientists to get a sense of how the disease will develop, since pigs grow up faster than we do, and allows them to test the impact of various drugs.

There will be a small group of pigs – perhaps five or so, according to Dhanu – for each child taking part in the study, all of which will bear exactly the same mutation on the NF gene as the child they represent. At present, there is no cure for NF1 and no way to treat the underlying disease itself. There are just drugs that combat the symptoms: behavioural changes and autism, tumours, cardiovascular disease and insomnia, as well as pain and eye problems.

'Mason's been introduced to a number of medications, stacking. He's on about a half-dozen different types,' says Chuck. 'Are they doing what they should? It's terrible. So we have this discussion on a daily basis, my wife and I: about just stopping all of Mason's meds. Are they actually doing anything or not? He has behavioural issues. Could the medicine be doing that? The disease varies so much from one person to another that it's hard to draw any general conclusions. Currently, the kiddos are the guinea pigs.'

In the first round, five children are to have their own pigs, and when I meet the team in June 2018 they're about to complete the last batch of paperwork before they can start creating their avatars. The first pigs are scheduled to be born in late 2018. But Mason won't be represented in the first group.

'No, no, no. That would be a conflict of interest. It wouldn't be the right decision,' says Chuck. It may seem odd that Mason isn't part of the group, but the fact that his father is involved in the project raises objections. Instead, a doctor, together with Dhanu, selects the children who are to take part. Mason may be included in a group later on, as the project develops.

Experiments of this nature, involving precise individual genetic modification of so many pigs, would have been impossible just a few years ago, for both financial and practical reasons. One of the benefits CRISPR brings may be the chance of a better life for Mason.

* * *

Jen Meudt is in charge of the practical aspects of gene-editing the pigs. She showers at the same time as me and shows me around among the animals. It's clear that she's a frequent visitor: she fondles the pigs' snouts, adjusts a sty door, checks equipment and food containers. There are about a hundred pigs here, 20 of which are involved in the study in one way or another. Many of them are sows from which eggs are removed for use in the research. The sows also function as surrogate mothers to gene-edited piglets.

Getting all the conditions required for genetic modification has been a time-consuming and complicated process. The gene-editing technique – using CRISPR to split the NF gene at precisely the right point – has been the least difficult aspect. Most of the effort has gone into understanding the sex lives and fertility of pigs.

Apparently there's no really effective way to check whether a sow's in heat, ovulating or ready to mate. You can't take a hormonal sample or have her urinate on a stick. Instead, the scientists have to line the sows up and parade a boar past them a few times so they can smell him.

Then they push down on the sows' hindquarters. A sow that's ready to mate adopts a particular position, tensing her muscles in a way that shows she can cope with a 500-kilo boar mounting her. Jen examines each sow's posture and checks how long the effect lasts. The sows that become egg donors and surrogate mothers also have to have fertility cycles that are in sync with each other. The sow that is to receive the donor egg must be exactly 24 hours behind the donor for everything to work.

Months of work culminate in a short window of opportunity once two matching sows are found. Once the scientists have extracted the eggs from the sows and fertilized them with sperm, the really complicated part begins. They have to complete the gene-editing procedure within four hours, before the fertilized egg starts to divide. These four hours can fall at any time – in a holiday period, in the middle of a weekend, at Christmas. If the process is carried out too late, the growing pig embryo won't have the required modification in all its cells.

The process of gene-editing is carried out manually. Scientists use a minute pipette under a microscope to inject a liquid containing the necessary chemicals together with the gene scissors and the instructions on the exact location for the snip. Then all they can do is wait and see. The embryos are implanted in a surrogate mother who will bear a litter of piglets, hopefully with just the modifications the scientists want.

When you talk about gene technology, the medicine of the future, it's easy to focus on labs, white coats, robots pipetting precise amounts of clear liquid. In theory, it all seems sterile and complication-free. But it's worth remembering how limited a picture that is. Much research involves blood, excrement and sperm, one after the other. Massive bodies lying on operating tables, delicate cells to be transferred from one body to another. Complex procedures and thousands of tiny components that have to match – and in the midst of it all, a bunch of inquisitive pigs and a sick little boy.

Given the tricky ethical balancing act all this entails, the scientists have started the Avatar project slightly more slowly than was originally planned.

'We didn't want to move fast without thinking through the ethics,' says Dhanu. 'We had to also figure out whether we were going to be blind to who the patient is. And how do you break the blinding, if you discover something that you need to report back to the patient?'

Chuck adds: 'One of the ethical examples is: what if we do mirror a child, and then that swine dies? Or that group of swine dies? How do you report that?'

What makes this project special compared with others is that the pigs are personalized: the sole purpose of each pig is to help a specific child to survive. So I ask how they see that – are we making pigs a bit too like us humans?

'You're wondering if we're on the way to blurring the lines, making something linked to humanity, are you?' asks Dhanu. 'All we're doing is altering certain molecular structures on particular genes. Sometimes the lay public seems to think that something becomes human if you change particular details in cells and the genome. I think that's very far from the truth: being human is far beyond that. Just adding a few molecular details doesn't turn a pig into a human.'

This project vividly illustrates the opportunities that gene technology has opened up: a move towards the individual and the specific, and away from the conventional standard medicines taken by everyone with a particular disorder. The next stage is to use these new genetic techniques to try to find full cures for NF1, instead of constantly treating its symptoms the way we do today. Scientists hope to be able to treat children with gene therapy that will either correct their genetic anomalies or enable their cells to ignore these anomalies. One of the problems is that such treatment might need to begin in the womb, as the NF gene affects so many things at an early stage of development.

Chuck hopes more parents will opt to get themselves and their children genetically tested as early as possible, enabling them to fight the disease more effectively. We're in the lab, among all the gear the team uses for fast DNA analyses, to identify our inner secrets as cheaply and efficiently as possible. While we're discussing all this amazing progress, I ask if he would like to see human embryos being gene-edited to prevent NF1, so no more children would be born with the disease. Hesitating, he mumbles for a moment before finally replying:

'The only thing that bothers me is the fact that things are moving so fast. We're moving so quickly right now. One of the things that worries me is we're jumping right into the clinical realm. But on the other side, I appreciate it too – because it gives me hope.'

10

A Huge Spider's Web

To understand how disorders like NF1 can be so complex, and why we can't simply mutate away our irritating propensity to catch a cold every February, we need to take a closer look at how exactly our genes work. Do you remember the question about why all human eyes are not a Mendelian blue or brown? We're on our way to finding an answer now.

When scientists began reading and interpreting genetic material, first that of microbes, then that of animals and humans, they were surprised at how much of it *wasn't* composed of genes. Most of it turned out to be something else. In fact, genes account for only about 2 per cent (about four centimetres) of the two metres of genetic material located in every human cell.[1] So what is the rest of that material? Much of it determines which genes are switched on, and which are switched off. Since each gene is an instruction for making a protein, it's clear that different types of cell must have different sets of active genes. A cardiac muscle cell, a skin cell and a nerve cell in the human brain all contain the same chromosomes and genome, but they have very different functions because each cell type has different active genes.

Certain chunks of DNA ensure that the chromosomes can

make copies of themselves (replicate) when a cell is about to divide, and help control the process so that each new cell receives the right number of copies of each chromosome. Other sections of our DNA, however – once known as 'junk DNA' – are still hotly debated by scientists. What exactly is this DNA for? How much of it is significant? What happens if you modify it? Little by little, science is gradually discovering what the different bits of it are for and how they function. There has been some progress, and each year brings new findings that reveal what different parts of this DNA do. But it will be a long time before scientists decide to try modifying them. For the time being, nearly all aspects of genetic technology, and all experiments involving CRISPR, focus on the genes or the tiny chunks of DNA that control them, not on the other parts of the genome. So we'll stay with the genes here.

The other startling discovery when the human genome was first sequenced was that we humans only have about 20,000 genes.[2] Twenty thousand minute chunks of DNA that determine which proteins our cells produce. That may sound like a lot, but it's considerably fewer than expected. Everything we are, everything our bodies do, is determined by these 20,000 tiny sets of protein instructions – and on the instructions the cell receives on how to use them. It's a bit like having a cupboard containing 20,000 ingredients you can use to make any recipe in the whole wide world. Just as with cooking, it's not just a question of what ingredients you have, but above all of how you combine them.

A gene is essentially a code for a protein. There are variants, of course, sophisticated processes within cells that enable different versions of a protein to be produced on the basis of the same gene. It's rather like a baker who makes her cakes lighter or heavier depending on how long she beats the mixture.[3] But that's also one of the major difficulties with using genetic technology to modify genes: if we aren't aware of all the different variants and special cases, we may accidentally get quite unexpected outcomes.

When the world's scientists first began to understand genes, and later on the link between them and what happens within the body, they hoped and expected that all this would be quite straightforward and easily comprehensible. If we could just manage to map our genes precisely enough, we'd identify a gene matching every trait. A gene for being overweight, a gene for heart problems, one for dementia, one for musical talent, one for schizophrenia and so on: more wide-ranging and complex than blue eyes and Gregor Mendel's peas, but still readily comprehensible and manageable. They got it wrong.

Of course, geneticists already knew that Mendel's theories didn't apply to certain traits, height being one of them. Humans aren't binary in this respect – either short or tall. We don't fall into discrete categories like Mendel's peas. Rather, all humans are of a similar height, some being shorter and others taller, while most are clustered in the middle. Few people are shorter than one and a half metres, or taller than two metres. The reason for this is that a person's height depends on a number of genes. A combination of genes determines how tall a person grows, along with other factors like nutrition and health, of course. But the link between genes and environment is a sidetrack we won't explore right now. It'll come later.

Scientists were quick to discover that complex traits, like height, intelligence and potential susceptibility to cardiovascular disease or dementia, were linked to more genes than simpler traits, such as eye or hair colour. As soon as they were unable to draw neat inheritance diagrams, they assumed that must imply that several genes were involved. In the late 1990s, scientists speculated that certain highly complex traits might be governed by as many as 20 or so genes – a startling idea. But again they were wrong.

For in the mid 2000s, scientists got a tremendous shock when, thanks to the mapping of the human genome and increasingly powerful computers, complex calculations could be performed to find out which genes collectively determine particular traits. They could scarcely believe the findings. This wasn't a matter of 20 genes, or 50, or even 100. In fact, scientists have now identified about 700 genes affecting human height. But the problem is that these genes account for less than one fifth of variation in height. So 700 genes are involved, yet their collective impact is quite limited. What possible explanation could there be?[4]

Although our environment clearly affects how tall we grow, it's heredity that accounts for most of it: tall parents have tall children. About 80 per cent of variation in people's height is determined by our genome.[5] Yet – even though scientists have identified so many genes linked to height – most of that 80 per cent remains unaccounted for. Presumably it is controlled by even more genes, each having even less individual influence on how much we measure in our stocking feet. This shows clearly that even when a trait is obviously hereditary and genetically determined – as height is – that doesn't mean that it's governed by a small number of genes. In contrast to Mendel's neat and tidy peas, we're looking at a curry in which each of dozens of spices makes a small contribution to the flavour and cannot be separated from all the others.

On top of this, scientists have found other tiny chunks of DNA that affect human height by determining which genes are active and which are not. To return to our kitchen metaphor, they are instructions about which ingredients, and what amounts of those ingredients, are to be taken out of the kitchen cupboard. A total of 100,000 of these tiny chunks – in addition to our genes – affect the height we reach. Moreover, this is just the DNA that has an influence significant enough to show up in computer models, which are still fairly crude. In the future, we are bound to discover more genes and more of the DNA that issues instructions.

The sheer number of genes involved raises another issue. Given that we have only 20,000 of them, 700 genes can't be concerned solely with height. That implies that each of those genes, and probably the 100,000 chunks of DNA that issue instructions, affect many other physical attributes and processes unrelated to height. It's as if the same ginger that gave extra piquancy to our curry now adds a spicy touch to a boeuf bourguignon, somewhere else entirely. This explains why there are some odd links: for instance, it seems that a gene which affects the risk of developing breast cancer also determines whether your ear wax is hard or soft.[6] We're looking at a tangled web that's pretty hard to disentangle, to put it mildly.[7]

Scientists have proposed that in some cases what determines

how short or how intelligent we are goes beyond a large number of genes: it may actually be the entire genome, every single bit of our DNA.

But hang on, you may now be thinking, there are people who are exceptionally short or enormously tall. That has to do with their genes, so surely it's genes that determine height?

That's partly true. There are various mutations and genetic defects that can make a person extremely short or extremely tall. Many of them are well known and have been researched in depth. In these cases, however, the defects of a single gene are so extreme that they affect the whole body: for example, a person's body never stops secreting growth hormone, or their skeleton stops growing far too early. There are genes that can clearly have a huge impact on height if they're defective, but in the majority of people, who don't have such mutations, they affect height no more than any other genes.

This is how it works in many cases. When everything is working normally, each individual gene has a tiny effect on the body – but a seriously defective gene can have a huge impact. Scientists and doctors today know of some 6,000 diseases and medical conditions caused by a single anomaly in one of the body's genes. The genes affected can be dominant or recessive, and some are linked to the sex chromosomes, so that an anomaly affects one sex only – mostly males – as in the case of the haemophilia that afflicted many European royal families over the course of history.

This is what inspires much of the hope we place in CRISPR and gene technology. If a disease or condition is caused by a single mutation, a single anomaly, it looks far more likely that it can be remedied by modifying that particular gene.

Most of the diseases that afflict us are complex, being caused by more than one gene and by the interaction of our genetic makeup with the environment. But even if gene technology can't eradicate them, there's hope nonetheless. In many cases, a mutation or variation can strongly affect the way the disease develops, and this is where gene-editing could make a difference, even if it didn't eliminate the risk of getting the disease.

What makes this more difficult is all the complex attributes that are affected by hundreds of genes and thousands of other pieces of DNA. This complexity means that we need to try to disentangle the various linkages to identify the locations where we can really make a difference, cure diseases and save lives.

This is where the metaphor of the kitchen spice cupboard rather breaks down. The same applies to the computer-code metaphor that many people use in attempts to describe how our genetic material works. Unlike food and computers, genetic material isn't put together or constructed through external agency: there's no cook or programmer to plan and structure it. A random process, combined with natural selection, has caused the formation of certain genes and the disappearance of others. And that, in turn, means that nothing in genetics is neat, simple or even particularly logical.

It's quite hard for us to grasp how different our genetic material is from the artefacts we've made ourselves, as nearly everything we can conceive of has been constructed by humans. The best point of comparison might actually be an ecosystem, such as a rainforest or a coral reef: a system in which many different living things, both plants and animals, interact and combine to form a whole. Totally unexpected links can arise within ecosystems, too. For example, the extinction of a type of ant could make it harder for a particular species of bird to survive, because the ant prevented insect pests from devouring the bird's preferred fruit, and so on. Yet it's impossible to be sure in advance what will happen if that particular ant dies out. Similarly, it's very difficult to identify and understand the various links between different genes.

Picture a hubristic future boffin bent on creating ultra-tall humans in a secret underground laboratory. That scientist can't simply tweak the odd gene, but will have to tinker with many different parts of the human genome. Even so, there's no way to be sure of gaining more than an extra half-centimetre. If the experiment does produce giants, there's a risk they'll face other problems, if some modification or other also happens to affect muscle growth or eyelash length.

This is something that emerged clearly in He Jiankui's experiment with Lulu and Nana. The gene he edited affects many processes in the body as well as the risk of HIV infection, and scientists are still discovering new links and connections. Nobody yet knows how the two girls' bodies will react in the course of a lifetime with the new gene variant; as a result of many different processes, their cells will have a slightly different composition. Thousands of dishes will be 'seasoned' slightly differently, simply because a single condiment in the spice cupboard has been swapped around.

As scientists identify more and more genes that affect different traits, it's also becoming clearer that the genetic code and all these interconnections are rather too complex for us humans to be able to disentangle and understand. Rather, many geneticists hope that artificial intelligence (AI) can help us analyse this gigantic spider's web – particularly the AI used to identify connections.[8] Masses of data, such as images of different kinds of animals, are fed into a program, together with basic instructions that resemble our nerve cells. The program then has to learn which images show cats, for example. What makes these systems so smart is that the programs themselves learn a way of differentiating between a dog and a cat. Such systems have got better, cheaper and faster over the last few years and can now be found more or less everywhere. Their many and varied applications include self-driving vehicles, automatic online translation, share analysis and facial recognition on Facebook, to name but a few.

Scientists and business are now trying to see whether the same methods can be applied to identifying dangerous mutations, links between genes and diseases, or new medical drugs tailored to the genomes of individual patients. AI can analyse thousands of genes and thousands of patients in parallel; it can search for needles in haystacks far greater than those we humans can cope with. One example is a large-scale US research project on motor neurone disease (amyotrophic lateral sclerosis, ALS), which seeks to link all the data from a group of patients so that an IT system can identify new connections, improving understanding of the disease and, hopefully, the treatment available. Many commentators believe this kind of intelligent computer program offers the only way of developing an in-depth understanding of our genes: a means of clarifying even more of the complexity that was first discovered by Gregor Mendel.

Perhaps AI will be able to help us clear up the question of eye colour. Because that irritating teenager who questioned Gregor Mendel's inheritance diagram was absolutely right. There isn't just one gene that determines our eye colour: reality is far more complex, and we still don't know exactly how it works. There are a few studies according to which at least 16 different genes affect eye colour.[9] Although the difference between blue and brown eyes mostly follows Mendel's rules, the other genes involved determine whether blue eyes are light blue, or a shade of grey or green, and whether brown eyes are dark or light, amber or nearly black, and everything in between. The colour of the human eye is also determined to some extent by the internal structure of the eye, which affects how light is reflected. And that structure depends on a different group of genes which haven't yet been completely identified.

So even the simplest example isn't really simple. Genetics just isn't as straightforward as it looks at first sight. There are layers upon layers of complexity, layers upon layers of confusion and nuances. Is this making you feel dizzy yet? It has that effect on me. But it's important to have a sense of just how complicated our genetic makeup is if we're to grasp the potential of gene technology.

We will probably be able to cure many genetic diseases and reduce the risks of various other disorders in which a handful of genes play a major role. But even if we manage to analyse this complex tangle using AI, countless traits will still be controlled by so many genes that it won't be possible to affect them much using CRISPR or any other method. That's one reason that our dreams of transforming ourselves or our progeny are so complicated in real life.

Although gene technology is a powerful tool which opens up entirely new possibilities, that doesn't mean that we can tinker with every single one of our traits. Genes aren't well-defined switches in an orderly grid, but a vast skein of interlinked threads like an endless ball of yarn. The mere fact that a trait is hereditary doesn't mean that we can control it through gene technology.

11

A Modern-Day Horoscope

I don't read thrillers. Being sensitive and easily scared, I've no desire to immerse myself in human depravity, even if fictional. However, I've had to study a series of violent crimes so hideous that I doubt whether even a thriller writer could have invented them. In 1970s and 1980s California, the same atrocity recurred again and again. A man would break into houses and apartments at night and pull a gun on the residents. His victims were women living alone or couples. He'd wake them abruptly and tie them up, together with any children in the house. Sometimes he'd pile crockery on his captives and threaten to kill everyone in the household if he heard the dishes clink. He raped the women, often repeatedly, and stayed in their homes for long periods. He'd help himself to food and beer, search through drawers and sometimes steal small items before finally disappearing. Sometimes he murdered his victims, sometimes he left them bound and bloodied.

Between 1974 and 1986 he broke into over a hundred homes, raped over 50 women and killed at least 13 people. Despite intensive police operations, the criminal couldn't be found. There were plenty of DNA samples from the crime scenes, but no match could be found with any of the suspects. In the end the crimes stopped, but the perpetrator, 'the Golden State Killer', remained at large – until 24 April 2018, the day on which 72-year-old Joseph James DeAngelo was finally arrested.[1]

The turning point came thanks to genetic tests for use at home. Tens of thousands of people worldwide have spat into test tubes and sent them off to various companies to find out more about their genes, their origins, the risk of contracting various diseases and the chances of locating unknown relatives. Many genome-sequencing companies also offer their customers the option of uploading their results to huge public databases, so they can identify relatives and find out more about their genealogy. The murderer wasn't as daft as to send in a sample himself, but that wasn't necessary.

With crime-scene DNA samples, help from volunteers and the support of family history researcher Barbara Rae-Venter, the police managed to identify ten distant relatives of the murderer. They linked together the various lineages. As they already knew the murderer's sex and approximate age, they were ultimately able to narrow the field down to just one suspect. They also matched the DNA samples with others taken from food packaging in the murderer's rubbish bin before arresting him. Joseph James DeAngelo has now been sentenced to life imprisonment with no possibility of parole.

The Swedish police recently used the same method to identify a man who raped an eight-year-old girl in 1995. The child was hospitalized for a while after this extremely brutal and vicious crime, but no one could be arrested until a change in the law enabled the police to identify relatives of the perpetrator in their database. Thanks to this change in the law, there's hope that more cold cases will be solved.[2]

With growing numbers of people taking genetic tests, more and more information is available in the databases of genome-sequencing companies, making it easier to identify almost anyone through their DNA. It's enough for some distant relative to have taken a test for a DNA sample from a crime scene to be linked with a suspect. A study carried out in 2018 showed that over half of all white Americans could already be tracked down and identified in this way.[3]

'In a few years it's really going to be everyone,' geneticist Yaniv Erlich has told *Science*. He says it'll be possible to identify anyone anywhere because someone who's sufficiently closely related will have had a home gene test.[4]

The first time the human genome was mapped, it cost nearly 3 billion US dollars. Just a few years later, the cost was a million dollars. Today you can get your genome fully sequenced for a few thousand dollars, perhaps even less by the time you read this. Full sequencing means that the genetic code is analysed in its entirety, even though about 5 per cent of the genome remains unmapped by scientists. This is because actual genes are only a small proportion of all genetic material, and certain parts of the genetic code can't yet be accurately read using today's technology. This may be because the same series of genetic 'letters' is repeated again and again. Inevitably, it's hard to be sure how many repetitions there are in the chromosome.

For just a few dozen dollars, you can order a relatively simple genetic test that will give you information about everything from your family tree to your risk of contracting various illnesses. The very first companies to offer gene testing for genealogical purposes appeared back in 2001, since when a variety of tests have been developed and hundreds of DNA companies have been set up. So far, over 30 million people have sent off their DNA to one of these. The biggest market player is 23andMe, only a small proportion of whose business involves genealogy. Most customers use it to check their risk of getting various diseases and to seek other lifestyle guidance based on genetic analysis.[5]

The tests all start in more or less the same way. You get a little kit, generally a test tube you have to spit into and a swab to rub inside your mouth and collect cells. Not many cells are needed for a genetic analysis.

Mapping the whole of a person's genome remains a costly and complicated business. That's why most tests rely on identifying genetic markers. These are tiny chunks of DNA which either manifest particularly large variations, meaning that there are very likely to be differences between people, or which have certain variants associated with a higher risk of particular diseases. Different tests are designed to identify different markers, ranging from a handful to a few thousand.

A decade or so ago, I, my father and my paternal grandmother took a swab from the inside of our cheeks. We wanted to find out what routes our ancestors had taken on the journey from Africa to southern Sweden. The genetic test we took didn't analyse the whole genome, just two types of DNA: the Y chromosome and mitochondrial DNA. Mitochondrial DNA is the genetic material outside the cell nucleus, which accompanies the egg, but not the sperm, when a new embryo is formed. (See Chapter 2.) That means that mitochondrial DNA is passed on in a direct line from mother to daughter, and that it can be traced all the way back to the woman who was the foremother of all the women alive today. Similarly, the Y chromosome is passed on in a direct line from father to son, and can be traced back to the man who was the forefather of all men living today. Of course, that doesn't mean our ancient foremother and forefather were the only people alive at the time, or even that they lived at the same time or met each other. It was just that their offspring – among the earliest humans – happened to reproduce successfully, producing daughters or sons who also managed to reproduce and multiply throughout history.

Mitochondria and Y chromosomes provide a means of tracing mass migrations because they don't mix in every generation. In contrast, all chromosomes but the Y chromosome are a mix of those you've inherited from each parent. One of your children can inherit the gene variant given to you by your mother, while another can inherit the gene variant passed down to you by your father. But the mitochondria and the Y chromosomes remain unadulterated. That doesn't mean nothing happens to them: they are subject to mutations and changes like the rest of your genetic material. And these changes can pinpoint when new variants emerge and spread over the globe. Conversely, huge numbers of Y chromosomes and mitochondria have disappeared in the course of history. I can take my own family as a simple illustration: I have no brothers, and nor does my father. My paternal grandfather had a brother, but he remained childless. That means that the variant of the Y chromosome that originated with my paternal great-grandfather has died out now, even though there are plenty of female relatives passing on the rest of his genes. And if I don't have any daughters, the mitochondrial lineage that began with my maternal great-grandmother will come to an end, as neither I, my mother nor my maternal grandmother have had any sisters.

Research has shown that these two types of lineage, Y chromosomes and mitochondrial DNA, have sometimes gone very different ways. This is largely a matter of chance, but there are other reasons for the disparity. Sometimes it reflects differences in the ways men and women have moved around the world. We know that Iceland, for example, was settled mainly by men from Norway, Vikings, and by women from the north of Ireland who were presumably taken there as slaves.[6] For at least the last 10,000 years, armies on the move, consisting essentially of men, have had a major impact on genetic change. Southern India provides one example. A group of people there claim to be of Macedonian ancestry, descendants of soldiers from the army of Alexander the Great who stayed on and had children with women already living in the region.

Power is another significant factor affecting the dissemination of individuals' genes. In a number of studies, geneticists have identified specific variants of Y chromosomes with a more widespread and rapid distribution than the norm. The most credible explanation is that they came from men who wielded great power. A study of 800 Irishmen identified a variant of the Y chromosome that was especially widespread in the north-west of Ireland. That area had been dominated for nearly six centuries by a clan claiming descent from a warlord called Niall Noígíallach (Niall of the Nine Hostages). Very little is known about Niall, and many historians have written him off as a mythological figure. However, research shows that the families claiming to belong to his lineage do have a common ancestor, a man living in the fifth century CE who founded an extremely powerful dynasty – presumably Niall himself. Between 2 and 3 million men alive today have inherited Niall's Y chromosome.[7]

The same pattern applies to two other historical characters: the Mongol emperor Genghis Khan and Giocangga, founder of China's Qing dynasty. In both cases, scientists have identified a particular variant of a Y chromosome which is more widespread than usual and which emerged during the lifetime of these historical personages. Clearly their genes were widely disseminated because these powerful men had many offspring themselves, but equally because their sons, grandsons and great-grandsons had power and wealth and thus more chance of having many children who survived into adulthood.

Of course, a perfectly ordinary man might also enjoy similar biological success just by chance, but it's unlikely. Scientists say this sort of effect is typical of regions that have had a strong power structure, with dominance passing from father to son. In theory, a woman's genetic heritage could also be passed on down the generations if power went from mother to daughter. However, it would take far longer for her mitochondria to become widespread, given the limits to the number of children she, her daughters and her granddaughters could bear in a lifetime.

It's now quite a long time since I took my genetic test, and many new types of test have since become available. There are huge variations in what they offer and how they work. One quite common option is to find out how much of your lineage comes from particular parts of the world: you might, for instance, be 30 per cent Swedish, 20 per cent British, 10 per cent Spanish, and so on. Such tests have earned strong criticism from scientists because the genetic makeup of historical populations in some countries is not very well documented.[8] So most of these tests are based on comparing your genes with those of a small sample of the present-day population of given regions or countries, not with the historical population. Major population movements, or pure chance in sampling, can make it look as if your origins are very different from what they actually are. Journalist Kristen V. Brown has written about a genealogy test she took which claimed that she had genetic roots in Italy, when she knows full well that her family has a Syrian background.

Since it remains expensive and tricky to sequence the genome in its entirety, most tests analyse just a limited number of locations. The selection of markers is very important. A test that focuses on certain markers (particular parts of the genome) may give a very different result from one based on comparing other markers. Many studies have shown that the same person can get very different results from different gene-test companies – even identical twins with identical genes sometimes get different results from the same outfit.

There are success stories too: adoptees reunited with members of their biological family, contacts and friendships between relatives who didn't even know of each other's existence before testing. Gene testing has been used to help bigoted far-right extremists grasp that they, too, have roots in other parts of the world. It's easy to be seduced by the notion that you can understand yourself and your story by digging down into your genes and chromosomes.

* * *

A new kind of gene test for use at home has emerged that has nothing to do with genealogy or geographical origins. Instead of looking into your history, the idea is to predict your future, using tests that show whether you risk developing any of a number of diseases or a range of other health problems.

In May 2013, the actor and producer Angelina Jolie wrote a frank and personal piece in the *New York Times* about why she'd opted for a double mastectomy several years before her fortieth birthday. She had watched her mother battle breast cancer and die of ovarian cancer at only 56. Her decision was radical, but logical on many levels. Jolie carries a gene variant called BRCA1 which, together with her family's history of breast cancer, gave her an 87 per cent chance of developing the disease. After the operation, she runs less than a 5 per cent risk of getting breast cancer.[9]

In her op-ed, Jolie urges all women who have the opportunity – especially those with a family history of breast cancer – to take a genetic test to find out if they carry genes that could put them at a higher risk. It's well known that mutations in two genes called BRCA1 and BRCA2 can significantly increase the risk of breast or ovarian cancer. Mutations can be passed on, making certain families particularly vulnerable. And these mutations aren't uncommon: in the United States, an estimated one in every 400 people has a mutation in one of the two. In Sweden, roughly 2 per cent of all breast cancer patients carry such a mutation. Men can also have the mutation, which increases their risk of developing many different kinds of cancer.

If you're worried you may have one of these variants, you may wish to get advice and help from a healthcare provider. But there are also several gene tests available that allow you to check for yourself whether you run a higher risk than average. They work in a similar way to genealogical tests: you send in a small cell sample and the gene-testing companies don't sequence the whole genome, but examine just those locations associated with breast cancer or other diseases. In the case of certain mutations, such as those affecting the two BRCA genes, the answer is very clear-cut. In others, however, it's more complicated.

We tend to think of DNA and our genes as a manual, a kind of software that tells us how we function. But it's not that simple. Certain genetic variants have a clear link with a particular disease. In some cases, such a mutation invariably results in disease, while in others it can increase the risk very significantly – as with breast cancer. But such associations are exceptions, extreme cases. Generally speaking, mutations with such negative effects are swiftly eliminated through evolution. Most people who develop breast cancer don't have a mutation in a BRCA gene. All those who get the disease despite not having that particular mutation are victims of a combination of factors: genetic makeup, lifestyle and a big dose of chance.

There are, of course, other genes that affect the risk of developing breast cancer, some of which scientists have identified, while others are still unknown. Just as there are over 700 genes affecting our height, most of our traits – from intelligence to the probability of having a heart attack – are controlled by hundreds, or sometimes hundreds of thousands of bits of DNA.

Many of the world's leading scientists are working to solve this puzzle. Even more geneticists are trying to find out what links exist, which is harder than you might think. The simplest way is to use lab animals, mostly mice or rats. This makes it easy for researchers to single out a gene they think significant and knock it out with CRISPR, after which they wait to see what happens. Alternatively, they can feed the same diet to two groups of mice with different variants of the same gene, put them on the same exercise programme or expose them to the same risks, and see whether that makes any difference. Since we share many of our genes with rodents, the results of such experiments are a useful first step. But such methods can only reveal the broad structures, not the fine detail of how they work.

Researching genes in humans is trickier. But now that gene tests are becoming cheaper and more common, it's possible to check whether a particular mutation is more usual among people with a given disorder. That was exactly how the link between breast cancer and the BRCA genes was discovered. From 1974 to 1990, geneticist Mary-Claire King studied families with a higher than usual incidence of breast cancer, searching for a possible genetic explanation.[10] At the time she embarked on her study, nobody thought cancer could be hereditary, and many wrote off her efforts as pointless. She herself has admitted in interviews that she often wondered whether her field of research was a blind alley. She tested and rejected over 150 different gene variants before finally discovering the one she named BRCA1, and was the first to show that the risk of developing cancer could be hereditary. A few years later, she and her research team identified BRCA2. The method she developed has enabled scientists to identify more links between genes and diseases. Although today's methods for analysing and sequencing genes are far better than in the 1980s, this nonetheless gives an idea of how much time and effort is required, even in the case of mutations that genuinely increase the risk of disease.

Another way to try to establish which genes are associated with given traits is to conduct experiments involving groups of people with different variants of the same gene, to see whether the results differ. Many such studies are carried out to see whether different variants of a given gene determine how hard or easy it is to lose weight, for instance. There are also many studies in progress into whether it is genetic differences that make certain medicines effective for some individuals, but ineffective for others. The problem with such studies is that they are expensive and difficult to carry out, so the experiments involved are often short-term, with small groups of participants. This means their results are less reliable than they might be.

* * *

Our genes lay the foundations for who we are. But we've only just begun to piece them together. Looking at our genes as a gigantic jigsaw puzzle, we've found quite a few corner and side pieces – those with a strong, clearly identifiable impact. We've also assembled a few bits here and there in the middle, but there's still a huge amount we don't understand, and many pieces of the puzzle remain in the shadows.

Things get even trickier when this knowledge is pressed into the service of tests designed for use by curious or anxious people. The companies that produce gene tests for the consumer market bundle together lots of different studies – some reliable, others less so – and base the tests they offer on their results.

The hundreds of different kinds you can buy test for all sorts of diseases and zero in on different locations in the genome to search for links. Many of these tests have come in for criticism because the companies selling them aren't upfront about how uncertain the results are and draw far-reaching conclusions based on a very few test results.[11] There are many examples of this. For instance, one company advises some of its customers to drink nearly a litre of apple juice a day to lose weight. This advice comes from a single study in which 68 obese men drank apple juice every day for four weeks.[12] The results showed that men with a particular gene variant lost some body fat, while the others didn't.

That's an interesting result that warrants further investigation – first and foremost to see whether other scientists get the same result, or whether chance happened to play a trick on the research team in this instance. But it's a pretty weak basis for advising people what to eat and drink. Examples of how a single study can determine gene-testing companies' recommendations and risk assessments are legion. In the scientific community, however, there's a consensus that you need to keep an open mind and an enquiring attitude when you've only got one study to go on – and that you absolutely mustn't take the results for gospel truth.

There are now some rules on the claims gene-testing companies can make, what tests they can offer and what conclusions they can draw in their recommendations. For instance, the US Food and Drug Administration (FDA) slammed 23andMe for their tests on the risks of developing various diseases. Customers used to receive information about more than 250 diseases; now the FDA reckons that only a handful of results are reliable enough to quote. Yet the law varies from one country to another, and this remains a largely unregulated market.

Many tests also offer answers to what you might call lifestyle issues: the best way to eat, exercise or sleep, depending on your genetic makeup. Other tests claim to tell you how creative you are, or whether you're stress-resistant or more empathetic than other people – without specifying what that might mean in practice.

In tests like these, all the scientific challenges – interactions between genes, major knowledge gaps and the limitations of the studies available – coincide with a very different issue. And that is the incredible plasticity of the human body and brain, with their ability to adapt continuously to circumstances, the environment, the things that happen to the individual and the way that individual is raised. Various gene variants have been linked to behaviours such as addiction, risk-taking, creativity, and so on. Yet our genes merely provide a set of underlying preconditions: one and the same characteristic may prove either helpful or harmful to an individual, depending on circumstances. Risk-taking may be bad for someone growing up in a social milieu conducive to criminal behaviour, but helpful to a potential Nobel-winning scientist. Moreover, even if you carry a gene that makes you a bit more of a risk-taker, it will be affected by the unique composition of all your other genes – and that's even before we take environmental factors into account. That risk-taking gene might never have any perceptible impact on your life if it's counteracted by some other gene.

In practice, it's still impossible to say all that much about a person's behaviour, mental health or personality based on their genes. There are a few exceptions, such as variants that make the carrier more likely to suffer from conditions like schizophrenia. And there are genes that seem to be able to increase or reduce that probability slightly. Yet such variants are just a small part of the mix of heredity, environment and chance in which our lives play out.

Despite this, the companies concerned are keen to draw attention to the enormous potential of their tests. For example, the CEO of Veritas Genetics – one among many DNA testing companies – said in an interview that they hope to create a 'Netflix for genetics'. And the co-founder of Helix, another such business, is on record as saying: 'The same way people use data to determine which movie to see or which restaurant to eat at, people will one day use their own DNA data to help guide everyday experiences.'[13]

This is where the problems with these tests start to become really clear. To give an example, parents can order a test for their young children and find out, thanks to a few gene variants, whether they'll be good at music, find it harder to learn to read, shine in maths or show a talent for languages. Clearly, parents might look at results like these and bring their children up slightly differently. They might be more lax about reading lessons: 'That's really hard for you – take it easy.' Or nag them a bit more about music practice: 'You're gifted, it's just you don't make enough effort.'

In the same way, it's easy to say to yourself: 'There's no point starting to run, my genes say I'd be better off doing weights.' Or: 'I don't think I'll bother with that French course – I'm rubbish at languages anyway.'

On top of that, reports are now emerging that some employers want their employees to take gene tests – just as others require personality tests – to assess where they'll fit best within the organization. Many geneticists have called such tests completely unscientific. In Britain and some other countries, scientists and doctors have called for a ban on DNA testing at home. One of the risks they point out is that these tests are often misleading: for instance, many women taking the test for BRCA genes have been wrongly informed that they carry dangerous mutations, while others who really do have the mutations have been told the opposite.[14]

*　*　*

There are also tests that are just plain silly, but quite entertaining: ones that use your genetic makeup to tell you what superpower you'd have if you were a superhero, which wines you'll enjoy or what kind of skincare will suit you best. There's even a dating site to match you – on the basis of your genes – with potential soulmates.[15]

There's nothing wrong with reading your horoscope just for fun. I know that chance formations of stars and planetary movements have no power to influence my life and can't affect the person I am or determine the decisions I should take. The idea is ridiculous. Yet I read my horoscope quite often, especially the slightly longer ones that appear at the beginning of each month. It's a bit of fun, not something I take seriously, though it can sometimes give me the nudge I need. 'It's particularly important to focus on friendships this month.' *Hmmm... It's been quite a while since I last invited friends round for dinner, hasn't it – about time I organized something...*

I can get something out of horoscopes without believing in them, benefit from a momentary break for reflection without allowing myself to be manipulated or tempted to splash out on a customized horoscope based on my personal details. Obviously, DNA tests focusing on exercise, stress management, dating, diet, sleep, and so on can be used in the same way – as a bit of fun, a starting point for reflecting on what you need in your life, an inspiration for changes that might do you some good. You can pick out what appeals to you and forget about the rest. But the problem with DNA tests is their superficially scientific appearance, the fact that they seem to come from an objective and truthful source. And the gene-testing companies tend to market them in exactly that way, no matter whether the results are scientifically meaningful or not.

It's not easy to tell which parts of the tests are based on reliable science and which ones are just a kind of horoscope, with no basis in reality.

Where gene-testing differs from horoscopes, however, is that your genes are more interesting to third parties than your horoscope is – look at the findings used to identify the Golden State Killer. Some insurance companies plan to offer lower premiums to people who have their genes tested to check their risk of developing certain diseases. Information from people's gene tests also ends up in other places. The companies that carry out these tests save all the information they collect and often sell it on to others. As with many services in the new digital age, you aren't the customer, but the product. Google is free to use because it sells information about you – and it's no coincidence that Google is one of the main investors in the biggest DNA testing company, 23andMe. Just as with Google, gene tests are as cheap as they are because information about your genome is saved and can be sold on to other companies. 23andMe, for example, sells its data on to GlaxoSmithKline, the pharmaceutical giant. At the moment, most of the firms that purchase such information are pharma companies, but there's nothing to stop the data being sold on to firms of a very different kind, or to prevent it being used for other purposes.[16]

It's not only individuals who are tempted by the promise of genetic testing. Several countries are now starting to deploy the technique in the hope of identifying promising athletes to bring home international gold medals. China and Uzbekistan have started genetic testing programmes, and the risks of gene doping are the subject of lively discussion in several sports.[17]

While the big pharma companies hold out the promise of a better future, there's another side to this science as well. In September 2018 I spent a few weeks in China gathering material for this book. It was at about this time that the world woke up to the Chinese authorities' surveillance and coercion of the Uighurs, an ethnic group in western China. A large proportion of the Uighur population – as many as a million people – have been put into 're-education camps', without having been accused of any crime. Uighurs are not permitted to pray, to wear veils or to grow beards in accordance with their traditions. Families are being broken up and people are disappearing. The aim is to destroy all aspects of the region's culture and to break any resistance to the regime. A video filmed by a drone shows thousands of blindfolded prisoners, with their hands tied behind their backs, being shipped off by rail to prison camps in other parts of China.

One aspect that might not seem immediately significant is that the Chinese authorities are apparently collecting genetic material from essentially the entire Uighur population, according to a report by Human Rights Watch. Several million Uighurs have been ordered to spit into test tubes as part of mandatory health checks on children and adults alike. This has been done with a strong element of coercion, and the people concerned haven't been told what the samples are to be used for. Given the number of DNA samples collected, the authorities can trace people, track down individuals who have moved to other cities and, for example, link them to DNA evidence associated with crime. The term 'crime' can include protesting against the actions of the Chinese authorities and having undesirable contacts outside China. Collecting genetic data on the population is the next logical step for a police state.

The Chinese authorities can track people down using the methods applied by the California police to trace the Golden State Killer. It has also emerged that they are looking for ways to find out whether individuals belong to minorities such as the Uighurs or Tibetans by examining their DNA. According to a patent application and other sources, the idea behind this is to infer 'geographical origin from the DNA of suspects at crime scenes'. The patent application also states that this type of documentation is necessary to fight terrorism, 'as such cases are becoming more and more difficult to solve'. These are vague circumlocutions for what basically comes down to registering people's ethnic origins.[18] This is being done although most scientists agree that a person's ethnic identity cannot be determined through genetic analysis. People have had sex and produced offspring regardless of borders throughout the history of humanity. Ethnic groups are seldom genetically distinct from their neighbours.

China's coercive DNA screening is one of the most sinister examples of how such tests can be applied. But the companies that provide gene-testing kits for use at home are also building up vast databases full of information which can be put to uses that would surprise the people who post their saliva to DNA testing labs.

12

A Place You Didn't Think Existed

The idea of transforming human beings has fascinated writers ever since Mary Shelley laid the foundations for science fiction in the early nineteenth century, and the theme recurs regularly in books, films and TV series. In the introduction to this book, I described how our view of genetic technology and our own bodies strikes a precarious balance between the fear that science will run amok (as in *Frankenstein*) and the hope that new discoveries could help create a better world (as in *Star Trek*). But there's a third narrative we shouldn't neglect.

Gattaca, a film made in 1997, is set in a 'not-too-distant future'.[1] More than 20 years on, that future is now so close at hand that we can reach out and grasp it – if we want to. The movie portrays a society in which people's biometric data is registered and most parents-to-be select their children's traits before conception. Though it's illegal in theory to discriminate against someone on genetic grounds, in practice it's impossible to get a prestigious job if your genes are less than flawless. The film tells the story of Vincent, who dreams of becoming an astronaut, but, having been conceived in the traditional way, is classed as genetically 'non-valid'. The society depicted in *Gattaca* is far-removed from the totalitarianism of *Brave New*

World; rather, it shows the logical conclusion of people opting freely for self-optimization.

One of the film's most intriguing aspects, if one compares it with *Frankenstein* or *Star Trek*, is that it doesn't depict science as inherently good or evil. Instead, the focus lies on the choices we make and the options before us.

For what fate awaits those who are not flawless in a world where everything can be transformed and enhanced, at least if you can afford it?

* * *

Jackie Leach Scully is a British academic.[2] I meet her in Boston at a conference about how gene technology can change the world. The impression she gives is of a powerful intellect coupled with an intense manner, a combination I see particularly often in British scientists. She refrains from overstatement and expresses herself lucidly.

Her scientific career began with research into what happens at the cellular level in breast cancer. Gradually she developed a growing interest in how the public views science and in who controls it. At the time of our meeting, she is a professor in bioethics at Newcastle University, and has just addressed the conference on the subject of ethical boundaries in genetic technology. She was an advisor when Britain decided to give the green light to mitochondrial replacement therapy and is assisting with an investigation of gene therapy as a whole at the time we meet.

'Unlike a lot of bioethicists [...] I'm not particularly interested in working out for myself what I happen to think the right answer is or the right thing to do. My main interest has always been in how people of different kinds come to those decisions: about where they're going to set limits, about how you decide when a technology is appropriate to be used,' she says.

These limits are currently being tested almost everywhere. In a Petri dish in the Salk Institute for Biological Studies,

California, there lies a tiny embryo barely four weeks old. It's just big enough for you to be able to see that it has the potential to develop into a living creature. You can see a round head and a curved back ending in a short tail. There are dark patches that would have become eyes if it had been able to continue growing. This is a mammal, but it would take an expert to tell what kind. In fact, this embryo would have grown into a piglet if it had been allowed to live – but a piglet of a very particular kind. For the cells within the embryonic piglet include human stem cells, and if the animal had been permitted to develop, about one in every 100,000 of its cells would have been human. The result would have been a mix of the two species.

Scientists inserted a few human cells early on in the embryo's development and let the different cells grow together. This isn't the first time researchers have tried combining cells in this way: it's been done with rats and mice, for instance, and it's even possible to control the development of specific organs so that they're almost entirely composed of cells from the other species. Rats and mice are similar, but not identical. Mice have gall bladders, an organ not found in rats. But rats with a few mouse cells can develop gall bladders and, if scientists make the necessary changes, other organs consisting almost entirely of mouse cells. Depending on the stage of embryonic development at which the new cells are inserted, such animals can grow into an equal or unequal mix of the two species. It also seems to be possible to control particular stages of embryonic development by genetic means. Juan Carlos Izpisua Belmonte, the scientist who led the experiment on the hybrid pigs, once grew an embryonic mouse limb on an embryonic chicken wing.

In interviews, Belmonte has said that research of this kind would take hundreds of years with conventional methods. 'But thanks to CRISPR, we can move quickly to many, many genes and modify them.'[3]

Scientists call these new creatures 'chimeras', after a mythical fire-breathing beast that was part lion, part goat and part serpent. Mixing cells is one way to blur the boundary between pigs and humans, but you can also go straight to the genetic level. Many other teams of scientists are experimenting with splicing selected human genes into pigs and gradually making them more like humans. Dhanu Shanmuganayagam, for instance, the scientist researching NF1 and trying to help Mason Konsitzke, is running another project looking into the possibility of inserting certain human genes into pigs.

The aim is the same, whether the scientists concerned mix human cells in with pig cells or methodically splice in one gene after another: to find a solution to the huge shortage of healthy organs for sick people. Tens of thousands of people die each year because they can't get a donor organ in time. More people are prepared to donate organs now and great strides have been made in preserving the organs of traffic accident victims. Yet the shortfall will almost certainly continue to take its toll, and it's also fuelling a horrific illegal trade in organs.[4]

Transplanting donor organs is always a tricky matter. It would work best if the organ came from the patient's own body, like skin transplanted from one part of the body to another. In such cases, the individual's immune system recognizes the cells and accepts them. Getting the body to accept an organ from another person is harder: medication is needed to stop it rejecting the foreign cells. Since the early twentieth century, scientists have also been trying to replace human organs by their counterparts from animals. This has proven incredibly difficult: the only such transplants that really work involve the cornea, or, in a few cases, parts of organs.[5] But now that gene technology is developing so fast, it may eventually be possible to rear a pig with bespoke organs matched to a specific individual: a walking, grunting spare-part kit.

Pigs with spare parts for particular individuals are just one example of all the possibilities that have appeared in the wake of the CRISPR revolution: possibilities that could have a profound impact on our society if the current phase of infatuation with gene technology fulfils its promise.

In the words of Juan Carlos Izpisua Belmonte: 'It could change our own evolution, our own species.'

But not all such experiments are designed to alleviate

human suffering. Su Bing, for instance, tried to blur the boundary between humans and monkeys.

Su Bing spliced a gene apparently associated with human brain development into 11 rhesus monkeys. Five of them survived, and their brains seem to have developed rather differently from the norm: they performed slightly better than others on a cognitive test. But these results are weak and inconclusive, and other scientists have observed that the only conclusions one can draw are that the experiment was as badly conducted as it was unethical. In an earlier experiment, another team of scientists took a gene that appears to be linked with human speech and inserted it into mice. The mice squeaked more often and performed marginally better on certain tests, but that was all.

More recently, Su Bing transferred another human gene to monkeys – a gene that emerged at about the same time as the ancient hominin Lucy. Some scientists believe it may have been decisive in differentiating us from other animals. Su Bing rejected the criticism directed at him, which he believed to be rooted in stereotypes about research in China.

'Exploring the genetic mechanism of human brain evolution is a major issue in the natural sciences, and we will continue our exploration,' he said to CNN. He brushes off criticisms of his research as unethical, stating that his team abides by China's rules on animal experiments.[6]

* * *

As far as Jackie Leach Scully is concerned, the essential question is how new medicine and gene technology will change the way we look at disability and difference in general.

'A lot of this has to do with disability, with trying to either prevent it or ameliorate it,' she says. 'So it has a lot to say about our attitudes to disability and to difference. And I think that's a particular issue with genome editing.'

She collects her thoughts momentarily, putting her explanation into clear, precise words before she continues.

'Much of the new genetic technology has been promoted and justified with the explanation that it will help prevent some people having some devastating genetic conditions and diseases and disabilities. And I think we'd all agree that that's a good thing. The problem emerges when you start moving away from the clearly devastating diseases that nobody in their right mind would want to live with, and you start following a kind of moving boundary to conditions that are perhaps more or less possible to live with, perhaps begin to edge into the territory of "This is just a difference". And genetics isn't going to tell us where those boundaries are.'

We need to bear in mind that the dividing line between someone with a disorder and someone who is healthy but different in some way isn't laid down by science alone. There are political and social aspects to such issues. How tall or short do you have to be before you're classed as having an abnormality, how fat or thin, how light or dark? How well do you need to be able to hear, see, speak, think? What we see here is that different countries and cultures choose to define the boundaries differently, to draw different lines in the sand.

The shifting and renegotiation of norms and definitions is more than just a theoretical issue facing bioethicists and scientific theorists. Jackie Leach Scully points out that these shifts immediately raise a multitude of questions, questions that are relevant both in Scandinavia and in the UK. For instance, which phenomena are classed as being so serious that sufferers are entitled to particular benefits and support? Which conditions will expose sufferers to discrimination, hate crime or other forms of hostility? What will happen to children born with genetic anomalies, mutations and diseases in a future society where disability is increasingly rare? Will people with disabilities still have the right to support if their parents could have gene-edited or aborted them?

'The boundaries between disabled and non-disabled have moved backwards and forwards. These aren't actually new issues,' she says; but she acknowledges that the new gene technology has made these concerns still more pressing.

Genetic technology also has the potential to change what

a human being is, to redefine the limits of the human and to determine what traits we have.[7] This potential goes beyond just curing diseases and medical conditions; it could, in fact, involve taking the first steps towards enhancing humans or transforming us into something entirely different. He Jiankui's gene-editing of the twin sisters Lulu and Nana lies on the cusp of such change. It is a first small step towards an enhanced human better adapted to a new era and freed from the genetic shackles that bind us today. But what are those shackles?

Jackie Leach Scully herself has a serious hearing impairment and refers to herself as deaf. It's easy enough to chat in a quiet hotel room, but at the conference, where the noise level is high, Jackie is assisted by a woman who transcribes everything for her benefit. Her impaired hearing is the result of a childhood illness, not a genetic anomaly. But the field she works in means she's constantly being asked whether she'd like to see a genetic cure.

'No, I wouldn't,' she says. 'My life is perfectly OK and if it were to change now I probably wouldn't be able to cope. I wear a hearing aid, and every time it gets upgraded it takes me an effort to get used to the new thing. A change in my life now would actually not be a particular advantage to me.'

Many people find it hard to take that on board, she says with a chuckle: that there are advantages in hearing less well than others, and that it's not nearly such a big problem as most hearing people think.

'There's a great deal of fear and ignorance about different kinds of disability,' she continues, 'and an assumption that you would always want to get rid of them. To my mind my hearing impairment is irritating.' Her tinnitus is a much bigger problem, she says, but no one asks if she'd like a cure for that – it's not viewed as being as serious as deafness.

There's a big difference between allowing individuals (such as would-be parents) a choice and the norms we establish collectively as a society. Society lays down guidelines; establishes the tests to be performed during pregnancy; decides whether it is acceptable to abort embryos with genetic mutations, whether couples who risk passing on genetic diseases should receive IVF treatment, and so on. But there are interactions between individual choices and social norms. One clear example is that fewer children with Down's syndrome are being born in Sweden, as would-be parents increasingly opt for a termination if amniocentesis reveals chromosomal anomalies.

One of the ethical risks of gene technology is that it has no ethical costs. A couple who have opted for IVF treatment have to select the embryo to be implanted in the woman's womb in any case. Selecting one that has no genetic anomalies could easily become the reasonable and logical choice, as could correcting a minor genetic detail with the help of gene technology.

'There are a lot of what we would today call disabilities which are perfectly compatible with a perfectly OK quality of life, given a few changes in the way society supports people with that particular difference,' says Jackie Leach Scully.

The problem is that we tend to talk about disabilities as if all of them had the same implications for the people concerned. When you talk about using gene technology to cure diseases and medical conditions, you need to specify exactly what you mean, what condition you're referring to, as she points out several times during our conversation. For instance, are you talking about a disease so serious that a child born with it won't live for more than a couple of years, about Down's syndrome, or about deafness? Many people in the deaf community would say deafness isn't actually a disability at all, Jackie clarifies.

I ask whether the easiest answer isn't simply that the better we can make things, the better in general.

It's not that easy to define what's 'better', Jackie replies. Being on the short side, like me, she says that as a teenager she'd have given practically anything to be a little bit taller, whereas her height doesn't bother her now. She can even save money by buying children's clothes, she jokes. And she points out that all of us have 'a range of potential problems or things that are more or less advantageous in different circumstances'.

*　*　*

The scenarios people have proposed of what we might be able to achieve in years to come are ambitious, to put it mildly. While some are probably best viewed as thought experiments rather than as to-do lists, they give us an indication of what might be possible in future.[8]

Some scientists have raised the possibility of making all humans meat-intolerant through a mutation that would make us unable to break down certain substances contained in meat. Meat would then be poisonous to us – a phenomenon already known to exist, which can be brought on by certain infections and the bites of some species of ticks. Meat intolerance would cut our greenhouse-gas emissions, while at the same time enabling us to feed more people from the land under cultivation. Admittedly, it would be a very blunt instrument and a dictatorial way to tackle a meat industry that damages both the climate and human health.

Others point out that long space journeys are probably not feasible unless we have people who are genetically engineered to withstand more radiation and build up more muscle and bone mass under low gravitation, as well as having other special adaptations. Some think the best way to colonize Mars would be to adapt ourselves to the red planet, rather than to make Mars more like the Earth.

A group of scientists are even discussing changing our feelings and instincts. They want to re-engineer our genes to make us more empathetic and less status-driven, so as to boost international cooperation in the interests of solving environmental problems and global conflicts. Others speculate about turning us dark green so we can carry out photosynthesis and get some of our energy straight from the sun. Somehow I can't see that particular adaptation catching on in Sweden, but it's a nice idea.

And this isn't just about the human body. There's hope that the new gene technology can be used to protect and conserve wild fauna and ecosystems. Ten or so separate research teams are working to resurrect extinct animals, from the northern white rhino – of which a very few remain, but which now have no further hope of procreating by natural means – to mammoths, aurochs and other creatures that have been extinct for millennia.[9]

Many scientists have also begun to deploy genetic engineering to steer nature itself in a particular direction. During the Zika virus outbreak in Brazil (2015–16), a company sought to slow transmission by releasing hundreds of thousands of genetically modified mosquitoes that lacked the capacity to spread the virus. A following step, currently hotly debated by scientists, is to use gene drives to eradicate malarial mosquitoes, using a genetic mutation.

Mosquitoes, like humans, normally inherit half of their genes from each parent. However, a few years ago American scientists discovered a curious part of the genome that has the capacity to be transferred to all of a parent's offspring, not just half of them. No one knows for sure how this mechanism originated; it's rather like a self-copying system that isn't either intrinsically positive or intrinsically negative. But if you insert a gene immediately next to the gene drive, the latter will ensure that that gene is passed on to all of the individual's offspring. Scientists could insert a gene to ensure that all future progeny will be male, thereby wiping out all malarial mosquitoes within a few generations. The problem is that today's scientists don't really know yet what might happen if such a mechanism escaped into the natural environment. Would it result in the extinction of other species of mosquito if a different species were able to mate with the modified mosquitoes? Would it accidentally wipe out an entire ecosystem? Alternatively, perhaps the mosquitoes would develop some form of genetic resistance that would make the whole experiment pointless.

Genetic modification could radically change our lives if applied on an even larger scale and in different areas of agriculture. Possible applications include new crops with novel features, ranging from potatoes with a low glycaemic index (currently being developed by scientists in Skåne, Sweden) to hardier and more drought-resistant varieties of rice, cassava, cotton and our other staple crops.

Genetic modification of bacteria or yeast to boost the efficiency of medicinal drug production is another area in which CRISPR can be applied. Today most insulin is produced using genetically modified bacteria, whereas in the past it had to be sourced from the pancreases of pigs. It will soon be possible to produce most of the complex molecules we use in perfumes, vitamins, pharmaceuticals and makeup with the help of gene-edited bacteria or yeast cells cultivated in vats. There are plans to produce the components of milk, eggs and leather using such methods. New York's Museum of Modern Art boasts a T-shirt made of leather produced by genetically modified yeast cells, whose maker, a company called Modern Meadow, is now marketing similar garments under the Zoa label.[10]

Microbes can already be used to produce substances like petrol, diesel and aircraft fuel – though these processes aren't yet efficient enough to be a going concern. Genetically modified plastic-eating bacteria may provide the solution to the floating garbage patches that pollute our oceans. If they do, the next question will be how we're to stop the bacteria devouring our lunchboxes and electric cables while we're still using them.

One of the more unexpected uses CRISPR may be put to is tackling the ongoing antibiotic-resistance crisis. As we've developed new antibiotics, bacteria have also stepped up their capacity to develop new ways to survive, and they just aren't affected by antibiotics any more. This is an arms race which bacteria currently seem to be winning. They mutate and develop new forms of resistance faster than we can come up with new antibiotics. That's where CRISPR comes in. Although it originated with the defence system bacteria use against viruses, CRISPR can be used to modify the viruses themselves, improving their ability to kill the bacteria they infect. Various research teams are currently investigating how well that might work in practice. It wouldn't end the arms race between us and bacteria, but it would give us an extra weapon.[11]

The list of highly creative transformations and scary scenarios can be extended almost ad infinitum. It's startling to read about the things that are now imaginable, even though they're still far from feasible. Mars is unlikely ever to be settled by dark-green, peace-loving folk with a meat allergy – but the mere fact that we can now discuss the hypothetical steps which might enable that opens up the possibility of thousands of other transformations, great and small.

And this brings us back to the fundamental ethical questions: is it right to decide what kind of people we want to be in future? Or what kind of people we should have in our society? And just who gets to make such decisions anyway?

Answers to these questions have so far tended to be pretty general. Both scientists and politicians like to say there needs to be a broad social consensus and that they don't want to go against public opinion. Yet though the possibility of modifying humans has been discussed for 40 years – since the first genetically modified animals were born – there has been little likelihood of it actually happening. Up till now. And now these decisions have to be taken.

The problem is the divergence of views on genetic modification, says Jackie Leach Scully. That applies to all societies, with views ranging from the conviction that gene-editing must be stopped to the belief that it should be encouraged and preferably regulated as little as possible. She thinks the way different countries deal with it will vary widely. Another problem is that these issues are hard to discuss, given the complexity of the underlying science. There is a tendency to exaggerate what is actually feasible.

It looks at the moment as if most societies incline towards stringent regulation of genetic technology. They tend to see it as a technology that should only be used *in extremis*. Yet the risk this attitude entails is that many people may continue to suffer when their suffering could be avoided, while those with resources will seek treatment in countries with a different legal framework. There's also a risk that society might swing to the other extreme and ban people with certain mutations from having children unless they undergo gene-editing, or impose pre-implantation genetic diagnosis on them. Given Europe's shameful history of forced sterilizations, such a scenario seems unlikely to become the norm here. In view of China's historical one-child policy, however, it seems plausible to imagine that country taking such action.

Jackie Leach Scully points out that there will always be people with disabilities and that we need to create a society that works for them too.

Studies have already been carried out in European countries which show that parents, especially mothers, who choose not to abort foetuses with identifiable disabilities are sometimes treated with disdain and blamed for their choice. Such criticisms come from those who think they took the wrong decision and that no taxpayers' money should be spent on their children. There is a risk that this attitude may become even more widespread in a society with more highly developed gene technology: under such circumstances, people born with a disability and those who are injured or fall ill in the course of their lives would receive less support and be forced to live in a society that took less account of their needs. We risk creating a social climate like that portrayed in *Gattaca* – but we also have the choice of harnessing the potential of gene technology to avoid a dystopia of that kind.[12]

Epilogue

Do You Suffer from Freckles?

As a child I used to watch the well-known TV series about Pippi Longstocking. There was one scene that really got to me. Strolling around town, Pippi and her friends Annika and Tommy spot a large sign outside a beauty parlour. It says: 'Do you suffer from freckles?' Pippi marches in, goes right up to the shop assistant and declares, 'No! No, I don't suffer from freckles.' Somewhat perplexed, the woman replies, 'But my dear child, your whole face is covered in them!' Pippi doesn't lose her cool, but retorts, 'Yes, but I don't suffer from them – I like them!'

I seem to have inherited my freckles mostly from my maternal grandfather, who had so many that they overlapped. At the age I knew him, his hair wasn't red any longer. But Grandma told me that the first time she saw Grandpa, with his freckles and red hair, he was up a ladder, painting a house pastel pink. That must have been quite a sight. As well as having freckles, I have light skin, blue eyes and red hair. I'm sensitive to the sun, a trait I seem to have inherited from my father as much as from my maternal grandfather. No matter how long I spend in the sun, all that happens is that I get sunburned, go red and acquire even more freckles. I can't tan – a lesson I learned again and again, painfully, in my teenage years.

All these characteristics go hand in hand with a few others that could have an impact on my life. There are a number of studies indicating that red-haired people need more anaesthetic to make sure they are unconscious before surgery. Conversely, we react to lower doses of certain analgesics than non-redheads. My light skin produces vitamin D effectively. But perhaps the most significant thing is that my freckly skin puts me at a far higher risk of developing skin cancer than most people.

I'm happy with my freckles, my hair colour and my pale-pink complexion. These are fundamental parts of who I am, which I would never want to change. Even when I coloured my hair, I just picked even deeper shades of red.

While writing this book, I thought about whether the risk of skin cancer makes red hair such a negative trait that my parents should have eliminated it. Neither is a redhead, so if they'd chosen to select one embryo or to gene-edit their baby, they could probably have ruled out any mutations of the MC1R gene. It's variants of this gene that give people red hair, and I very probably carry one.[1]

So where is the dividing line between what's ethically defensible and what isn't, if we start eliminating genes that cause diseases and other problems? Certain choices are easier for physicians, ethicists, scientists and parents to make – such as the choice of trying to protect children from being born with diseases that are bound to kill them. Other issues are on the borderline, such as a significantly heightened risk of developing cancer, cardiovascular diseases or dementia early in life: diseases that multiply the risk of an early death. Then we come to the trickier cases, such as a higher probability of losing one's eyesight or hearing. Is it reasonable to discard or to gene-edit an embryo if the future child runs a heightened risk of impaired sight?

The mere idea of choosing not to have a red-haired child seems absurd. But where does the ethical dividing line run? And who gets to choose?

Moreover, there's also a fundamental existential risk if gene-editing and selecting the genes of future generations become options open to any of us, just as commonplace as making sure our children go to good schools or any of the other choices parents can make today to give their kids a head start. This isn't just about the risk of mistakes or off-target effects. The major risk we run is of becoming ultra-homogeneous and forfeiting the chance of being pleasantly surprised by our genes. If we just play it safe and opt for gene variants we know are favourable, we'll never change or develop. Nor will we ever discover any beneficial new traits that might otherwise fall to us by chance. We may succeed in avoiding much disease, death and suffering, but at the same time avoid having children who, though perhaps particularly vulnerable to developing cancer, might also change the world. In such a future, we can assume that geniuses like Stephen Hawking or Ada Lovelace would never be born in the first place.[2]

Conversely, it would be foolish to disregard the amazing potential of gene technology to cure and protect us from disease: in fact, it could even be used to make a few improvements. If I'd been born later, might I have been gene-edited to have freckles *and* better protection against the sun – or maybe even a fetching green complexion that would combine the ability to photosynthesize with the advantage of setting off my red hair? I wouldn't have minded at least one snip of the CRISPR scissors to avoid having to wear glasses. While there's much that seems disturbing in tomorrow's world of gene technology, I can't help hankering after all the amazing experiences that await us. Many business people, writers and intellectuals worldwide, including Steve Jobs and the historian Yuval Noah Harari, have called our time the century of biology. The twentieth century was the time when physics revolutionized the world with everything from electricity and the internal-combustion engine to nuclear weapons and space travel. In our new century, biology and gene technology are poised to bring about a revolution of equal magnitude.[3] We have acquired a set of the first basic tools we need to modify our genes. CRISPR could perhaps be compared to a bronze axe, as opposed to the stone tools we've had up to now. But we tend to forget that although we now have an amazingly powerful tool at our disposal, we've only just begun to understand our own genes. The mere fact that we've got ourselves a splendid new hammer doesn't make every problem a nail.

Our knowledge of genes and gene technology developed

at about the same time as computers and the Internet became part of our lives. That coincidence has had a major impact on the way we view our bodies. We talk about 'the genetic code' and about 'programming' cells or DNA to do new things. Biohackers take much of their inspiration from the computer-hacker movement and talk about gene technology setting off a similar revolution. This language suggests a radically new way of looking at the human body, implying that we're just a set of easily programmed codes, or a collection of interchangeable components amenable to regular upgrading and optimization.

The dream of optimizing humankind raises problems. By viewing ourselves as machines, we risk reducing ourselves to the merely mechanical: to things that are easy to define, devoid of complexity, every component of which can be fully understood because every component has been engineered. There are no uncharted, unexpected or unsolved mysteries. Yet our genetic material is an intricate web that we have only just begun to disentangle. Viewing ourselves as machines rather than animals could make us miss or fritter away the chance of the amazing journey of discovery before us. We risk narrowing our focus to mere nuts and bolts, when we could be exploring a mystery.[4]

Our understanding of what genes are, how they function and how they affect the rest of our biology remains extremely limited, rather like the first sketches of a world map made by early sixteenth-century Europeans after Columbus's return from his voyage to the Americas. That's why I think it's better to compare the way ahead to a voyage of discovery, rather than a computer code. A computer is a device built by humans, an artefact we're familiar with and understand in detail. That's not how biology works, even if the language of the DNA molecule resembles a programming language.

We still know very little about what happens inside our cells and the possible impact of tinkering with our genes. That's not necessarily an argument for inaction – rather the reverse. But it is a reason to be cautious and aware of our limitations, bearing in mind that we still have much to learn. We're deceiving ourselves if we claim to control the system already. This is just the start of an exhilarating adventure.[5]

Acknowledgements

When I started to write my first book, I was positive that the creative torments plaguing other writers wouldn't affect me. I was writing a work of popular science, after all, so I wasn't a 'real' author. Of course, I couldn't have been more mistaken.

Despite that experience, I embarked on this project with the same misconception: I was convinced that the difficulties inherent in writing a second book – which plague nearly all writers – wouldn't affect me. Again, I was mistaken: this book is the result of a fairly long and convoluted process.

That makes me even more grateful to all those who have helped and supported me throughout this project. At my Swedish publishing house, Natur & Kultur, it was incredibly useful to be able to discuss ideas with my publisher Lena Forssén. Throughout the process, she had more faith in both me and the book than I did myself. My editor, André Borchert, did an invaluable job of tidying up my language and clarifying my arguments. The book was greatly improved by Erik Kohlström's fine illustrations. And it would never have existed without the talented quintet of translator Fiona Graham, editor Christian Müller, copy-editor Alex Middleton, typesetter Ditte Løkkegaard and designer Sophie Burdess, who created the cover. It was also very reassuring to have researcher Magnus Lundgren's scientific assistance with fact-checking my manuscript.

I would like to thank all the scientists who gave their time to be interviewed, and to answer my questions and help me understand their projects and results. No scientist works in isolation, but to keep the book clear and readable I confined myself to naming only the key players in each project – and sometimes not even those. However, the projects I refer to obviously involved many more scientists. I was helped in my research by many people who are not mentioned in the book. The ethnologist Susanne Lundin gave me her time for a long and rewarding conversation at the start of the whole process, and I am very grateful for being able to benefit from her expert knowledge and ideas.

While I was given an incredible amount of help and support during this process, you can't live on love alone. So I would also like to thank the Längman Cultural Fund, the Helge Ax:son Johnson Foundation, the Swedish Authors' Fund and the Swedish Authors' Photocopying Fund for the grants that enabled me to travel and to work on this book.

The hardest aspect of being a writer is working alone, so it was a wonderful thing to spend time in 'Kreativator', which may be the world's best co-working space, and to lunch regularly with fellow writers in 'Skånska Skrivare'. But what I'm most grateful for is my friends' affection and help, which has included taking long tea or coffee breaks together to talk through issues arising in the book; reading through the manuscript; discussing, over lunch, how best to explain mitochondria; or – perhaps most important of all – sharing a cup of tea and a hug on the days when I suffered from self-doubt. You're all wonderful, and none of this would have been possible without you.

I also want to thank my family for steadfastly backing me in my belief that being a writer is a good idea, and that I'm doing the right thing by taking a leap into the unknown.

And finally, thank you, Tobias. I'm so glad to have you, and you make my life – and my books – so much better.

Glossary

Cells
All life forms (except viruses) are made up of cells. Unicellular (single-celled) organisms, such as bacteria, consist of just one cell, while the human body is composed of billions of cells. Our bodies contain many different cell types – from muscle cells to nerve cells – all of which have a nucleus containing a copy of our chromosomes.

Chromosomes
The genetic material inside the cell nucleus is divided among 46 chromosomes in 23 pairs. Each pair has one chromosome from each parent. (This doesn't apply to people with certain chromosomal anomalies.) A chromosome is a long DNA molecule. All human cells contain identical copies of all chromosomes, which are copied (replicated) each time a cell divides. All of a person's chromosomes put together are known as the 'genome'.

CRISPR-Cas9
CRISPR-Cas9 is a tool that enables us to edit (or modify) DNA very precisely. CRISPR was discovered in bacteria. The acronym stands for 'clustered regularly interspaced short palindromic repeats', meaning bits of genetic material that form short chunks of code which can be 'read' in both

directions. A chunk of CRISPR RNA, copied from DNA, can be used to control an enzyme (Cas9). This enzyme works like a pair of scissors, cutting through DNA. By designing chunks of CRISPR RNA, scientists can control exactly where CRISPR-Cas9 cuts through genetic material. This enables them to remove or add material, to dismantle the DNA molecule or to modify it.

DNA

DNA stands for 'deoxyribonucleic acid'. This is the molecule that makes up the genetic material of all life on earth (except for certain viruses; see *RNA*). The DNA molecule, which looks like a twisted ladder forming a double helix, lies in the nucleus of each cell. The cell can 'read' the DNA molecule in order to synthesize proteins (and for some other purposes).

Genes

The genetic material in our chromosomes serves a variety of purposes, and the DNA molecule has many different parts, rather like sections or chapters in a book. Genes collectively account for about 2 per cent of our DNA. Each gene codes for a protein. Proteins are versatile, making up hair and nails, hormones and the enzymes that regulate processes within the cell. We have about 20,000 genes, some of which can generate more than one protein, depending on how the proteins are put together within the cell.

The rest of our DNA (98 per cent of it) is not made up of genes, and is known as 'non-coding DNA'. It is a mixture of sequences of DNA that control a variety of processes, including which genes are active and how the DNA molecule is copied.

Gene technology / genetic technology

These expressions cover all the methods used to read the genetic code and modify genetic material. Genetic modification means altering the genetic code. This may involve inserting new genes, as in the case of the bacteria which produce insulin. Alternatively, it can involve editing existing genes, as He

Jiankui did with the twins Lulu and Nana. It can also refer to repairing genes with an anomaly resulting in a genetic disease, or deleting genes that cause problems.

Gene therapy
Treating a disorder by manipulating a patient's genetic material is known as 'gene therapy', a technique first applied to human patients in the 1980s. It normally involves trying to splice in genetic material in one of a variety of ways, to replace defective genes. The DNA molecule can be edited using CRISPR-Cas9 or by other methods; alternatively, RNA can be inserted into the cells to give them new instructions. One of the tricky aspects of gene therapy is delivering the new genetic material to the cells. Non-infectious viruses are often used for this purpose.

Genetic code
The genetic code contains all necessary information about how cells, the body and living organisms are put together. Both DNA and RNA are made up of substances that can be 'read' like letters in a long chain. The order of these letters determines how the cell builds proteins, and all other processes within the body.

The four letters are adenine (A), guanine (G), cytosine (C) and thymine (T).* In our DNA, the A and the T are always opposite each other, and the same applies to the G and the C. That's why they are referred to as 'base pairs'. The length of genes, chromosomes and DNA is measured in base pairs. Since all life on earth has the same genetic code, we know that everything alive today has the same origins.

A DNA analysis involves 'reading' the string of letters, either throughout the genome or in small segments of it.

GWAS
GWAS, which stands for 'genome-wide association study', refers to large-scale statistical studies in which scientists

---

* In RNA, thymine is replaced by uracil, U for short.

try to establish a link between genes and certain traits or disorders. This is done by comparing the genetic material of many different people. Many genes affect a wide variety of traits, so it can be difficult to establish the exact influence of a given gene.

RNA

RNA, ribonucleic acid, resembles a comb or half a ladder. It is a kind of genetic material used by cells to transfer information from DNA in the cell nucleus (nuclear DNA) to the rest of the body. There are many variants of RNA: tiny copies of the DNA sequences to be used by cells, the RNA that transports and build proteins in the cell, and RNA sequences that determine which DNA is to be copied.

RNA is more liable to mutate than DNA. That's why most of life on earth uses DNA to store information, and RNA to apply it. The exception is a group of viruses that have only RNA, but no DNA.

Sources and Further Reading

Here are my sources, along with plenty of in-depth information and further reading about all the subjects referred to in the book, plus a few digressions which didn't really fit into the main body of the text.

Many of my sources are only available online, or at least are far easier to find on the Internet than elsewhere, so the whole of this section is available on my website, with up-to-date links to all references. See www.kornfeldt.com/the-unnatural-selection-of-our-species/.

Introduction – A Time of Radical Change.

1   The official Nobel Prize website lists all the Nobel laureates for 2020: https://www.nobelprize.org/all-2020-nobel-prizes.

2   There is a fascinating book about the scientific background to *Frankenstein*: Kathryn Harkup's *Making the Monster: The Science behind Mary Shelley's Frankenstein* (London: Bloomsbury, 2018).

      The first experiments showing that electricity could make animal corpses move were conducted by Luigi Galvani in 1780. Jessica P. Johnson's article 'Animal electricity, circa 1781', describing these experiments, with contemporary illustrations, appeared in *The Scientist* on

28 September 2011: http://www.the-scientist.com/?articles.
view/articleNo/31078/title/Animal-Electricity-circa-1781/.

3   In 1818, Mary Shelley (as she was by then) published *Frankenstein; or, The Modern Prometheus* anonymously in London. Her name didn't appear on the novel until a translation was published in France in 1823.

4   The original series of *Star Trek* follows the crew of the USS *Enterprise* (NCC-1701), but the crews, the vessel itself and the stories have been constantly updated ever since. Everything (and almost certainly much more) you might ever want to know about the *Star Trek* universe can be found at https://intl.startrek.com.

5   The kiss between Captain James T. Kirk and communications officer Nyota Uhura was shown on American TV on 22 November 1968, in 'Plato's Stepchildren'. Though not the first kiss between a white American and an African American to be shown on TV, it had a huge impact at a time of great antagonism within the United States. Uhura is often highlighted as an icon and an early positive image of an African American woman with considerable power and gravitas. More on *Star Trek*'s iconic female characters can be found in Margaret Kingsbury's article '*Star Trek* created feminist icons in front of and behind the camera': https://intl.startrek.com/news/star-trek-created-feminist-icons-in-front-of-and-behind-the-camera.

6   Charles C. Mann's book *The Wizard and the Prophet* (New York: Knopf, 2018) focuses in depth on this conflict between different ways of looking at science and progress.

7   There is a great deal of research into how much healthier and more secure we are today. You can find more information about this at www.gapminder.org or in an article entitled 'Seven reasons why the world is improving' by Julius Probst, published on 11 January 2019 at https://www.bbc.com/future/article/20190111-seven-reasons-why-the-world-is-improving.

8   According to the World Health Organization (WHO), approximately 4.2 million people die each year from air pollution (https://www.who.int/airpollution/en/), while about 1.5 million die of diabetes (https://www.who.int/news-room/fact-sheets/detail/diabetes) and just under a

million people die of AIDS-related diseases (https://www.unaids.org/en/resources/fact-sheet).

9   Chapter 2 deals with the birth of Lulu and Nana in more detail. The first publication to report on the event was *MIT Technology Review*, in an article by Antonio Regalado published on 25 November 2018: 'Exclusive: Chinese scientists are creating CRISPR babies': https://www.technologyreview.com/s/612458/exclusive-chinese-scientists-are-creating-crispr-babies/.

10   My previous book, *The Re-Origin of Species*, was published by Scribe in 2018: https://scribepublications.com.au/books-authors/books/the-re-origin-of-species.

11   Louise Brown was the first person born thanks to in-vitro fertilization (IVF). Here, in her own words, is an account of the breakthrough and IVF: 'Louise Brown on 40 years of IVF: "I was the world's first IVF baby, this is my story"', *Independent* (25 July 2018): https://www.independent.co.uk/life-style/health-and-families/ivf-baby-louise-brown-story-test-tube-world-first-40th-anniversary-a8455956.html.

12   Jean Purdy received less recognition for her work, despite playing an essential part. Her role is beginning to be acknowledged more now. See, for instance, Iliana Magra, 'Three created a fertility revolution with I.V.F., but one, a woman, went unrecognized', *New York Times* (20 July 2019): https://www.nytimes.com/2019/06/10/world/europe/jean-purdy-ivf-plaque.html.

13   Comparisons were made between IVF and the atomic bomb. Feelings ran high and it was met with strong opposition. Here is an interesting article about the criticism once directed at IVF: Kat Eschner, 'In vitro fertilization was once as controversial as gene editing is today', *Smithsonian Magazine* (27 September 2017): https://www.smithsonianmag.com/smart-news/vitro-fertilization-was-once-controversial-cloning-today-180964989/.

14   Here is the press release announcing the award of the Nobel Prize in Medicine to Robert Edwards for his work in developing IVF: https://www.nobelprize.org/prizes/medicine/2010/press-release/.

15   Naturally, much has been written about Jesse Gelsinger and his tragic death, but one of the best accounts I have

read is 'Gene therapy emerges from disgrace to be the next big thing, again' by Carl Zimmer, published in *Wired* on 13 August 2013: https://www.wired.com/2013/08/the-fall-and-rise-of-gene-therapy-2/.

Another article about Jesse Gelsinger and the impact of his death on the development of gene therapy is Meir Rinde, 'The death of Jesse Gelsinger, 20 years later', Science History Institute [website] (4 June 2019): https://www.sciencehistory.org/distillations/the-death-of-jesse-gelsinger-20-years-later.

16   Ignaz Semmelweis was a fascinating person in many ways. Much has been written about him. See, for instance: Howard Markel, 'In 1850, Ignaz Semmelweis saved lives with three words: wash your hands', PBS [website] (15 May 2015): https://www.pbs.org/newshour/health/ignaz-semmelweis-doctor-prescribed-hand-washing.

17   More on the Vipeholm experiments: Sandee LaMotte, 'The Swedish cavity experiments', CNN [website] (30 October 2019): https://edition.cnn.com/2019/10/30/health/swedish-cavity-experiment-wellness/index.html.

1. A Modern Prometheus

1   The whole of He Jiankui's presentation and the subsequent question-and-answer session are reproduced on the website of the Second International Summit on Human Genome Editing: https://www.nationalacademies.org/gene-editing/2nd_summit/.

Both can also be found on YouTube in the video entitled '28 Nov 2018 – international summit on human genome editing – He Jiankui presentation and Q&A': https://www.youtube.com/watch?v=tLZufCrjrN0.

A transcription of the presentation and the Q&A session can be found at https://diyhpl.us/wiki/transcripts/human-genome-editing-summit/2018-hong-kong/jiankui-he-human-genome-editing/.

2   The five short YouTube videos that He Jiankui released before the summit can be found on 'The He Lab' channel at https://www.youtube.com/channel/UCn_Elifynj3LrubPKHXecwQ.

3   The CCR5 gene which He Jiankui tried to edit has a fascinating history. A. P. Galvani and J. Novembre give a scientific account of the gene and what is known about the variants that afford protection against HIV in their scientific article 'The evolutionary history of the CCR5-Δ32 HIV-resistance mutation', *Microbes and Infection* 7/2 (February 2005), pp. 302–9: https://www.sciencedirect.com/science/article/abs/pii/S1286457904003636.

4   Clearly it's difficult to know exactly how many cells there are in an adult human body, but the study most often cited refers to an average of 37.2 trillion (37,000 billion) cells. See Eva Bianconi et al., 'An estimation of the number of cells in the human body', *Annals of Human Biology* 40/6 (November–December 2013), pp. 463–71: https://pubmed.ncbi.nlm.nih.gov/23829164/.

5   Obviously, endless articles have been written about He Jiankui's experiment and its consequences, yet much remains unknown. Jon Cohen's excellent article 'The untold story of the "circle of trust" behind the world's first gene-edited babies' gives a good account. See *Science* (1 August 2019): https://www.sciencemag.org/news/2019/08/untold-story-circle-trust-behind-world-s-first-gene-edited-babies.

Another article on He Jiankui's experiment is David Cyranoski, 'The CRISPR-baby scandal: what's next for human gene-editing', *Nature* (26 February 2019): https://www.nature.com/articles/d41586-019-00673-1.

6   In June 2019 a scientific article was published (Xinzhu Wei, 'CCR5-Δ32 is deleterious in the homozygous state in humans', *Nature Medicine* 25) which appeared to show that double sets of the CCR5 variant that He Jiankui tried to create shorten people's lives. But in September 2019 it emerged that the study had serious methodological flaws. It has now been withdrawn by the scientists concerned.

The following is a news article on the problems involved: Rebecca Robbins, 'Major error undermines study suggesting change introduced in the CRISPR babies experiment shortens lives', Stat [website] (27 September 2019): https://www.statnews.com/2019/09/27/major-error-undermines-study-suggesting-change-introduced-in-the-crispr-babies-experiment-shortens-lives.

7   The 1997 Oviedo Convention can be read in its entirety on the website of the Council of Europe. See 'Convention for the protection of human rights and dignity of the human being with regard to the application of biology and medicine: convention on human rights and biomedicine': https://www.coe.int/en/web/conventions/full-list/-/conventions/treaty/164.

8   The First International Summit on Human Genome Editing was held in Washington DC in December 2015. Information about all the presentations and recordings from the summit may be found here: https://national-academies.org/gene-editing/Gene-Edit-Summit/.

'On Human Gene Editing: International Summit Statement' (the joint statement and agreement concluding the summit) may be found at https://www.nationalacademies.org/news/2015/12/on-human-gene-editing-international-summit-statement.

A quotation from the text: 'It would be irresponsible to proceed with any clinical use of germline editing unless and until (i) the relevant safety and efficacy issues have been resolved, based on appropriate understanding and balancing of risks, potential benefits, and alternatives, and (ii) there is broad societal consensus about the appropriateness of the proposed application. Moreover, any clinical use should proceed only under appropriate regulatory oversight. At present, these criteria have not been met for any proposed clinical use: the safety issues have not yet been adequately explored; the cases of most compelling benefit are limited; and many nations have legislative or regulatory bans on germline modification. However, as scientific knowledge advances and societal views evolve, the clinical use of germline editing should be revisited on a regular basis.'

9   The first scientific study to show that CRISPR-Cas9 could be used on human embryos was Puping Liang et al., 'CRISPR/Cas9-mediated gene editing in human tripronuclear zygotes', *Protein & Cell* 6/5 (May 2015), pp. 363–72: https://link.springer.com/article/10.1007/s13238-015-0153-5.

10   Tang Lichun's scientific article on the use of CRISPR-Cas9 to modify human embryos is here: L. Tang, Y. Zeng, H.

Du, et al., 'CRISPR/Cas9-mediated gene editing in human zygotes using Cas9 protein', *Molecular Genetics and Genomics* 292 (2017), pp. 525–33: https://link.springer.com/article/10.1007/s00438-017-1299-z.

11   A journal got hold of He Jiankui's draft of a scientific paper about the twin girls. The following article contains excerpts from the paper and scientists' comments: Antonio Regalado, 'China's CRISPR babies: read exclusive excerpts from the unseen original research', *MIT Technology Review* (3 December 2019): https://www.technologyreview.com/s/614764/chinas-crispr-babies-read-exclusive-excerpts-he-jiankui-paper/.

12   There is sure to be more news by the time you read this, but here is one of the first articles about Rebrikov: Antonio Regalado, 'A Russian scientist has threatened to make more CRISPR babies', *MIT Technology Review* (11 June 2019): https://www.technologyreview.com/f/613661/crispr-babies-gene-editing-scientist-china-russia-hiv/.

And here is an interview with one of the couples who want to take part in his experiment: Jon Cohen, 'Deaf couple may edit embryo's DNA to correct hearing mutation', *Science* (21 October 2019): https://www.sciencemag.org/news/2019/10/deaf-couple-may-edit-embryo-s-dna-correct-hearing-mutation.

13   The following article attempts to summarize the debate between scientists about whether, when and how we should modify human embryos. It contains a number of useful links to arguments on both sides of the debate, for those interested in further reading: Françoise Baylis and Marcy Darnovsky, 'Scientists disagree about the ethics and governance of human germline editing', Hastings Center [website] (17 January 2019): https://www.thehastingscenter.org/scientists-disagree-ethics-governance-human-germline-genome-editing/.

2. Build Your Own Baby

1   The badges worn by Korean women to show that they are pregnant were introduced to guarantee them a seat on public transport. A Korean website for expats provides

more details of the scheme: https://www.korea4expats.com/ article-pregnant-women-badge.html.

2   The news about South Korea's record low birth rate was reported in The Guardian and elsewhere: Benjamin Haas, 'South Korea's fertility rate set to hit record low of 0.96', Guardian (3 September 2018): https://www.theguardian.com/world/2018/sep/03/ south-koreas-fertility-rate-set-to-hit-record-low.

3   South Korea has embarked on an active pronatalist policy, but research indicates that certain aspects of the policy are harmful to women's reproductive health: Sunhye Kim, 'Reproductive technologies as population control: how pronatalist policies harm reproductive health in South Korea', Sexual and Reproductive Health Matters 27/2 (2019), pp. 6–12: https://www.tandfonline.com/doi/full/1 0.1080/26410397.2019.1610278.

4   One of the main reasons for continued population growth even in countries with a low birth rate is that the calculation is based on the number of children a woman has during her life. In other words, statisticians look at whether there are already large numbers of girls who will have children in the course of their lives. Existing children will have children during their lives, so the population will continue to rise for a few decades even after the number of live births per woman has fallen below replacement level (which means an average of 2.1 children per woman over a lifetime). An example of this is Japan: the number of live births per woman fell below 2.1 in 1974, but owing to the time lag described, the Japanese population didn't begin to fall until 2010, over 30 years later.

5   A report on reproductive tourism in South Korea: Ko Dong-Hwan, 'Korean clinic treats more foreigners with fertility problems', Korea Times (12 July 2018): http://www. koreatimes.co.kr/www/nation/2018/10/119_252097.html.

6   The CHA Fertility Center has a website (http://seoul. chamc.co.kr/en/) which provides more information about its activities. After my visit to the CHA Fertility Center, its American clinic was rocked by a scandal, a tragic case involving confusion between the embryos from three couples undergoing IVF treatment: 'PWCK: CHA Fertility Center

put couple through "a living hell" in having another woman give birth to our son', Bloomberg [website] (10 June 2019): https://www.bloomberg.com/press-releases/2019-07-10/pwck-cha-fertility-center-put-couple-through-a-living-hell-in-having-another-woman-give-birth-to-our-son.

7   In which country will the first clinic offer gene-editing to its customers? This remains an open question. A few articles have been written that try to provide an answer.

An account of legislation on gene-editing of embryos in a number of countries: Heidi Ledford, 'Where in the world could the first CRISPR baby be born?', *Nature* (13 October 2015): https://www.nature.com/news/where-in-the-world-could-the-first-crispr-baby-be-born-1.18542.

See also Guillermo Aquino-Jarquin, 'Tighten up Mexico's regulations on human gene editing', *Nature* (26 February 2019): https://www.nature.com/articles/d41586-019-00666-0.

8   Ishii Tetsuya has written a number of scientific articles on the risks of modifying embryos, the possible impact on society and the link with reproductive tourism. They include:

- 'Germline genome-editing research and its socioethical implications', *Trends in Molecular Medicine* 21/8 (August 2015), pp. 473–81: https://www.sciencedirect.com/science/article/abs/pii/S1471491415001070.
- 'Providing appropriate risk information on genome editing for patients', *Trends in Biotechnology* 34/2 (February 2016), pp. 86–90: https://www.sciencedirect.com/science/article/abs/pii/S0167779915002498.
- 'Germ line genome editing in clinics: the approaches, objectives and global society', *Briefings in Functional Genomics* 16/1 (January 2017), pp. 46–56: https://academic.oup.com/bfg/article/16/1/46/2962382.
- 'The ethics of creating genetically modified children using genome editing', *Current Opinion in Endocrinology, Diabetes and Obesity* 24/6 (December 2017), pp. 418–23: https://journals.lww.com/co-endocrinology/Abstract/2017/12000/The_ethics_of_creating_genetically_modified.7.aspx.

9   In addition, many genes that were once part of mitochondrial DNA have now become part of the DNA in our cell nucleus, our genome, making the link between us and our mitochondrial DNA even stronger.

10   Not only mitochondria, but also the green chloroplasts in plants have the same history. For more detail, I recommend W. Martin and M. Mentel, 'The origin of mitochondria', *Nature Education* 3/9 (2010), p. 58: https://www.nature.com/scitable/topicpage/the-origin-of-mitochondria-14232356.

11   The NHS website contains an introduction to mitochondrial disorders: https://mitochondrialdisease.nhs.uk/patient-area/what-mitochondrial-disease/.

12   Two good articles on mitochondrial diseases and donations:
- For those who read Swedish, the magazine *Forskning & Framsteg* published an article in 2015 about the significance of the new law in the UK for Swedish couples who want to have children but suffer from a mitochondrial disease: Per Snaprud, 'För första gången tillåts en behandling som går i arv', *Forskning & Framsteg* (8 April 2015): https://fof.se/tidning/2015/4/artikel/for-forsta-gangen-tillats-en-behandling-som-gar-i-arv.
- The *New York Times* published an article about the fact that several decades have passed since the first genetically modified children were born using donated mitochondria: Carl Zimmer, 'Genetically modified people are walking among us', *New York Times* (1 December 2018): https://www.nytimes.com/2018/12/01/sunday-review/crispr-china-babies-gene-editing.html.

13   See the article referred to in note 9 of this chapter, but also the following scientific article: Jacques Cohen et al., 'Birth of infant after transfer of anucleate donor oocyte cytoplasm into recipient eggs', *Lancet* 350/9072 (19 July 1997), pp. 186–7: https://www.thelancet.com/journals/lancet/article/PIIS0140-6736(05)62353-7/fulltext.

14   A news report on the Ukrainian clinic: Rob Stein, 'Clinic claims success in making babies with 3 parents' DNA', NPR [website] (6 June 2018): https://www.npr.org/sections/healthshots/2018/06/615909572/

inside-the-ukrainian-clinic-making-3-parent-babies-for-women-who-are-infertile.

Much has been written about the opportunities and risks surrounding donated mitochondria. One example is this popular-science article: Catherine Weiner, 'Mitochondrial transfer: the making of three-parent babies', Science in the News [website], 22 August 2018: http://sitn.hms.harvard.edu/flash/2018/mitochondrial-transfer-making-three-parent-babies/.

15   A report claims that IVF clinics in Asia will be worth almost US$14 billion by 2025. *The Economist* has a good account of how this market is developing:

- 'The fertility business is booming', *Economist* (8 August 2019): https://www.economist.com/business/2019/08/08/the-fertility-business-is-booming.
- See also 'Asia-Pacific IVF services market is expected to reach $13.66 bn through 2025', Allied Market Research [website] (12 June 2019): https://www.globenewswire.com/news-release/2019/06/12/1867647/0/en/Asia-Pacific-IVF-Services-Market-is-Expected-to-Reach-13-66-Bn-through-2025.html.

16   See Antonio Regalado, 'Disgraced CRISPR scientist had plans to start a designer-baby business', *MIT Technology Review* (1 August 2019): https://www.technologyreview.com/s/614051/crispr-baby-maker-explored-starting-a-business-in-designer-baby-tourism/.

17   The ethnologist Susanne Lundin has written an excellent book on the trade in human organs, including the trade in donor eggs and sperm: *Organs for Sale: An Ethnographic Examination of the International Organ Trade* (Basingstoke: Palgrave Macmillan, 2015).

18   A well-written account of the problems that arise with gene-editing becoming part of the reproductive industry: Rosa Castro, 'The next frontier in reproductive tourism? Genetic modification', The Conversation [website] (18 November 2016): http://theconversation.com/the-next-frontier-in-reproductive-tourism-genetic-modification-67132.

19   A scientific article on the mice with two mothers: Zhi-Kun Li et al., 'Generation of bimaternal and bipaternal mice from hypomethylated haploid ESCs with imprinting

region deletions', *Cell Stem Cell* 23/5 (1 November 2018), pp. 665–76: https://www.cell.com/cell-stem-cell/fulltext/S1934-5909(18)30441-7.

20 A scientific article summarizing what we know about intelligence, genes and heredity: Robert Plomin and Sophie von Stumm, 'The new genetics of intelligence', *Nature Reviews Genetics* 19 (2018), pp. 148–59: https://www.nature.com/articles/nrg.2017.104.

21 News article: Hannah Devlin, 'IVF couples could be able to choose the "smartest" embryo', *Guardian* (24 May 2019): https://www.theguardian.com/society/2019/may/24/ivf-couples-could-be-able-to-choose-the-smartest-embryo.

22 A scientific article about the genetic link between creativity and certain mental illnesses: Robert A. Power et al., 'Polygenic risk scores for schizophrenia and bipolar disorder predict creativity', *Nature Neuroscience* 18 (2015), pp. 953–5: https://www.nature.com/articles/nn.4040.

23 The joint director of the European Bioinformatics Institute, Ewan Birney, has blogged about why it is problematic to try to select embryos with 'good genes', given that our traits are affected by a range of different genes: 'Why embryo selection for polygenic traits is wrong' (26 May 2019): http://ewanbirney.com/2019/05/why-embryo-selection-for-polygenic-traits-is-wrong.html.

(A quotation: 'This is wrong on a number of levels; ethically it is wrong to make this decision as an independent laboratory without broad societal buy-in; scientifically it is wrong to imagine the ways we assess polygenic traits will translate into safe and effective embryo selection; for the specifics of IQ/ Educational attainment trait this trait is so complex [that] this is additionally unwise over and above any concerns.')

24 Here are two studies of public attitudes to gene-editing:
- George Gaskell et al., 'Public views on gene editing and its uses', *Nature Biotechnology* 35 (2017), pp. 1021–3: https://www.nature.com/articles/nbt.3958.
- Masato Uchiyama et al., 'Survey on the perception of germline genome editing among the general public in Japan,' *Journal of Human Genetics* 63 (2018), pp. 745–8: https://www.nature.com/articles/s10038-018-0430-2.

3. Biohackers

1   For more background, see Genspace's website (https://www.genspace.org/) and an article on DIY biology written by co-founders Ellen Jorgensen and Daniel Grushkin in 2011: 'Engage with, don't fear, community labs', *Nature Medicine* 17 (2011), p. 411: https://www.nature.com/articles/nm0411-411?proof=t.

2   A website called 'A Guide to DIY Bio' contains a wealth of information about DIY biology, from labs to courses and conferences to projects and active Facebook groups. It was updated in 2019. See https://thatmre.medium.com/a-guide-to-diybio-updated-2019-abd0956cdf74.

      See also https://sphere.diybio.org/, a website that collects information about labs, projects and so on.

3   Biotech Without Borders has a website at http://www.biotechwithoutborders.org/.

4   A video in which Josiah Zayner injects himself can be viewed at: https://www.buzzfeednews.com/article/stephaniemlee/this-biohacker-wants-to-edit-his-own-dna

      Zayner has a blog, Science, Art, Beauty, which includes a post from February 2017 on genetically engineering humans: http://www.josiahzayner.com/2017/02/.

5   Netflix has a website about the series *Unnatural Selection*: https://www.netflix.com/title/80208910. There are four programmes that look into different aspects of gene technology, including trying to edit your own genes.

6   Here are a few news stories about how various people before Josiah Zayner tried editing their own genes:

- On Liz Parrish: Dara Mohammadi and Nicola Davis, 'Can this woman cure ageing with gene therapy?', *Guardian* (24 July 2016): https://www.theguardian.com/science/2016/jul/24/elizabeth-parrish-gene-therapy-ageing.
- On Brian Hanley: Antonio Regalado, 'One man's quest to hack his own genes', *MIT Technology Review* (10 January 2017): https://www.technologyreview.com/s/603217/one-mans-quest-to-hack-his-own-genes/.
- On an attempt to cure lactose intolerance: David Grossman, 'Scientist painstakingly documents his

own DIY gene therapy', *Popular Mechanics* (14 February 2018): https://www.popularmechanics.com/science/health/a17804128/scientist-painstakingly-documents-his-own-diy-gene-therapy/.

7   A news article about Aaron Traywick, his experiment and his tragic death: Kirsten Brown, 'What does an infamous biohacker's death mean for the future of DIY science?', *Atlantic* (5 May 2018): https://www.theatlantic.com/science/archive/2018/05/aaron-traywick-death-ascendance-biomedical/559745/.

8   The company, called 'The Odin', has a website which tells you how to edit your own genes: http://www.the-odin.com/diyhumancrispr/.

9   Here's a comment from Josiah Zayner's public Facebook page from December 2018: 'Most people can't get access to these clinical trials and instead suffer and die. These gene therapies could be replicated for a few thousand dollars and made with the same purity.

      'With gene therapies there is very little chance of adverse events. However, it is illegal to provide this to people even under medical supervision. More than 1 million people in the world suffer from muscle diseases and the average life expectancy is 20–30 years. We need to do more to help these people. #dmd #musculardystrophy #biohacking #diygenetherapy #PiratingGeneTherapy.'

10   Here are two good interviews with Josiah Zayner (he declined to be interviewed for this book):
      - Tom Ireland, 'I want to help humans genetically modify themselves', *Guardian* (24 December 2017): https://www.theguardian.com/science/2017/dec/24/josiah-zayner-diy-gene-editing-therapy-crispr-interview.
      - Sarah Zhang, 'A biohacker regrets publicly injecting himself with CRISPR', *Atlantic* (February 2018): https://www.theatlantic.com/science/archive/2018/02/biohacking-stunts-crispr/553511/.

11   Here's a news article about the research and the ensuing debate: Kai Kupferschmidt, 'A paper showing how to make a smallpox cousin just got published. Critics wonder why',

Science (January 2018): https://www.sciencemag.org/
news/2018/01/paper-showing-how-make-smallpox-cousin-
just-got-published-critics-wonder-why.

12 Michael Schrage, 'Playing God in your basement',
Washington Post (31 January 1988): https://washingtonpost.
com/archive/opinions/1988/01/31/playing-god-in-your-
basement/618f174d-fc11-47b3-a8db-fae1b8340c67/.

13 Some more suggested reading for those interested in
the subject:
  - Eleonore Pauwels, 'The rise of citizen bioscience:
    is self-experimentation with gene editing techniques
    something we should herald as a new form of
    "permissionless" innovation?', *Scientific American*
    (5 January 2018): https://blogs.scientificamerican.
    com/observations/the-rise-of-citizen-bioscience/.
  - Marcy Danovsky, 'Hacking your own genes: a recipe
    for disaster', LeapsMag [website] (18 January 2018):
    https://leaps.org/hacking-genes-recipe-disaster/.
    This article highlights the similarities between DIY
    gene therapies and the problems that have arisen
    with stem cell therapy.

4. The Gene Race

1 Fifty years after Sputnik 1, in 2007, American National
Public Radio (NPR) made a series of programmes about
the satellite, its history and its impact: *Sputnik at 50:
Looking Back at the Space Race.* See https://www.npr.org/
series/14933218/sputnik-at-50-looking-back-at-the-space-
race?t=1573814938464.

2 Many commentators have called China's advances in
gene technology 'Sputnik events'. One such is André
Choulika in the following op-ed: 'The West is losing the
gene editing race. It needs to catch up', Stat [website] (29
October 2018): https://www.statnews.com/2018/10/29/
west-is-losing-gene-editing-race/.

3 Can China become a scientific superpower?
  - *The Economist* tried to answer this question in a long
    and fascinating article that's well worth reading and
    goes beyond gene technology. It also offers a good

account of the amount various countries and regions are investing in research: 'Can China become a scientific superpower?' (12 January 2019): https://www.economist.com/science-and-technology/2019/01/12/can-china-become-a-scientific-superpower.

- The lead article in the same number of *The Economist*, quoted in the text: 'How China could dominate science – should the world worry?' (12 January 2019): https://www.economist.com/leaders/2019/01/12/how-china-could-dominate-science.
- A slightly older news article in *Nature* which also aims to give an idea of how much China is spending on research: Richard Van Noorden, 'China by the numbers – research capacity has grown rapidly, and now quality is on the rise', *Nature* (22 June 2016): https://www.nature.com/news/china-by-the-numbers-1.20122.
- A good long article about how China is developing into a scientific superpower: Shellen Wu, 'China: how science made a superpower', *Nature* (1 October 2019): https://www.nature.com/articles/d41586-019-02937-2.

4 The Yunnan Key Laboratory of Primate Biomedical Research doesn't have a dedicated website. However, there are a number of good articles about the centre, including the following:

- Sarah Zhang, 'China is genetically engineering monkeys with brain disorders', *Atlantic* (June 2018): https://www.theatlantic.com/science/archive/2018/06/china-is-genetically-engineering-monkeys-with-brain-disorders/561866/.
- David Cyranoski, 'Monkey kingdom – China is positioning itself as a world leader in primate research', *Nature* (21 April 2016): https://www.nature.com/news/monkey-kingdom-1.19762.

5 The scientific paper Ji Weizhi referred to, describing how his team was the first to use CRISPR to genetically modify monkeys, is Yuyu Niu et al., 'Generation of gene-modified cynomolgus monkey via Cas9/RNA-mediated gene targeting in one-cell embryos', *Cell* 156/4 (13 February

2014), pp. 836–43: https://www.cell.com/cell/fulltext/S0092-8674(14)00079-8.

6   There are a number of good articles giving an overview of the feasibility of using monkeys as models for human diseases and the problems that entails, one of which is Kenya Sato and Erika Sasaki, 'Genetic engineering in nonhuman primates for human disease modeling', *Journal of Human Genetics* 63 (2018), pp. 125–31: https://www.nature.com/articles/s10038-017-0351-5.

7   Here is a scientific account of how monkeys are used to research age-related human diseases, such as Parkinson's, and the associated problems: Jean-Michel Verdier et al., 'Lessons from the analysis of nonhuman primates for understanding human aging and neurodegenerative diseases', *Frontiers in Neuroscience* (4 March 2015): https://www.frontiersin.org/articles/10.3389/fnins.2015.00064/full.

8   There have been many reports and studies on China's laboratory monkeys. See also 'Monkey kingdom' (note 4 in this chapter).

- Apoorva Mandavilli, 'Monkey business', *Nature Medicine* 12 (March 2006), pp. 266–7: https://www.nature.com/articles/nm0306-266.
- Baoguo Li et al., 'The primate extinction crisis in China: immediate challenges and a way forward', *Biodiversity and Conservation* 27/13 (November 2018), pp. 3301–27: https://link.springer.com/article/10.007/s10581-018-1614-y.
- Qingyong Ni at al., 'Conservation implications of primate trade in China over 18 years based on web news reports of confiscations', *PeerJ* 6/e6069 (6 December 2018): https://www.ncbi.nlm.nih.gov/pmc/articles/PMC6286804.
- Kathryn Bayne and Jianfei Wang, 'Oversight of animal research in China', in Javier Guillén, ed., *Laboratory Animals* (2nd edn; n.p.: Academic Press, 2018), pp. 243–66.
- J. Hagelin, 'Use of live nonhuman primates in research in Asia', *Journal of Postgraduate Medicine* 50/4 (2004): http://www.jpgmonline.com/article.

asp?issn=0022-3859;year=2004;volume=50;issue=4;spage=253;epage=256;aulast=hagelin.

9   Here are a few reports and articles concerning problems with academic freedom in contemporary China:

- Emilie Tran, 'China: zero tolerance for academic freedom', The Conversation [website] (17 October 2017): https://theconversation.com/china-zero-tolerance-for-academic-freedom-85200.
- 'Freedom in the world 2019', chapter on China, published by Freedomhouse: https://freedomhouse.org/country/china/freedom-world/2019.
- Human Rights Watch, 'World Report 2019: China events of 2018': https://www.hrw.org/world-report/2019/country-chapters/china-and-tibet.
- Scholars at Risk Network, *Obstacles to Excellence: Academic Freedom and China's Quest for World-Class Universities* (24 September 2019): https://www.scholarsatrisk.org/resources/obstacles-to-excellence-academic-freedom-chinas-quest-for-world-class-universities/.
- Jeffrey Mervis, 'When Europeans do science in China', *Science* (11 September 2019): https://www.sciencemag.org/news/2019/09/when-europeans-do-science-china.

10   It isn't clear how scientific progress in China will affect the country's regimented society, or vice versa. It's also difficult to assess how much freedom scientists have. Sociologist Joy Y. Zhang has researched China's science establishment. In this popular-science article, she describes how China has created a system that is vulnerable to falsification of research and to secrecy in the scientific community: 'Why Chinese science seems so secretive – and how it may be about to change', The Conversation [website] (24 January 2019): https://theconversation.com/why-chinese-science-seems-so-secretive-and-how-it-may-be-about-to-change-110326.

11   Much was written about Su Bing and his research when he first showed that he could splice human gene variants into monkeys, and Chapter 12 contains further references to his research. The following article is worth reading: Antonio Regalado, 'Chinese scientists have put human

brain genes in monkeys – and yes, they may be smarter',
MIT Technology Review (10 April 2019): https://www.
technologyreview.com/2019/04/10/136131/chinese-
scientists-have-put-human-brain-genes-in-monkeysand-
yes-they-may-be-smarter.

5. *Storm in a Bowl of Yoghurt*

1 Just like everything else in biology, the reality is rather
more complex. The microbes we commonly refer to as
bacteria fall into two groups: bacteria and archaea. These
are quite dissimilar, archaea being closer to eukaryotes
than genuine bacteria. Eukaryotes form a large and diverse
group. They include all multi-celled organisms: animals,
plants and fungi. We humans are eukaryotes, but the group
also includes all single-celled microbes other than bacteria
and archaea – a web of life so complex that it would take an
entire book to describe these organisms adequately. Three
well-known examples are amoebas, slipper animalcules
and the sea sparkle (*Noctiluca scintillans*), a luminescent
phytoplankton that lights up the west coast of Sweden in
late summer, as well as appearing in other places, including
the Norfolk coast in eastern England, the west coast of the
United States, the Somali coast and around Indonesia.
Every living organism on Earth falls into one of these three
categories. Viruses are somewhat separate from these three,
as technically they're not even really alive. They have no
cell of their own and are therefore not even counted as
unicellular organisms.

2 In the case of a person about whom so much has been
written as Alexander Fleming, it's almost impossible to
single out just one source. However, since he received the
Nobel Prize for his discovery (together with Ernst Boris
Chain and Howard Walter Florey), the short text about this
award at nobel-prize.org is a good place to start: https://
www.nobelprize.org/prizes/medicine/1945/summary.

3 A scientific account of the history of antibiotics and the
future challenges we face: Kyriacos C. Nicolaou and
Stephan Rigo, 'A brief history of antibiotics and select
advances in their synthesis', *Journal of Antibiotics* 71

(2018), pp. 15–184: https://www.nature.com/articles/ja201762.

4   An interesting podcast that explores how antibiotics have been used on animals and in food, particularly chicken: Cynthia Graber and Nicola Twilley, 'The birds and the bugs' [an episode of the Gastropod podcast] (15 August 2017): https://gastropod.com/the-birds-and-the-bugs/.

5   There are several summaries of all the discoveries that led to CRISPR becoming such a useful tool:

- The Broad Institute collects the discoveries and lists the main publications in the field: 'Crispr timeline': https://www.broadinstitute.org/what-broad/ areas-focus/project-spotlight/crispr-timeline.
- The website What Is Biotechnology? has a more detailed list starting in 1987, which is regularly updated: 'CRISPR-Cas9: timeline of key events': https://www.whatisbiotechnology.org/index.php/ timeline/science/CRISPR-Cas9.

I won't list all the scientific publications associated with this development. There is a partial list of some of the most important publications below:

- First discovery of CRISPR: Y. Ishino et al., 'Nucleotide sequence of the iap gene, responsible for alkaline phosphatase isozyme conversion in Escherichia coli, and identification of the gene product', *Journal of Bacteriology* 169/12 (December 1987), pp. 5429–33: https://www.ncbi.nlm.nih.gov/ pmc/articles/PMC213968/.
- The study of milk products which showed that CRISPR was an adaptive immune system: Rodolphe Barrangou et al., 'CRISPR provides acquired resistance against viruses in prokaryotes', *Science* 315/5819 (2007), pp. 1709–12: https://pubmed.ncbi. nlm.nih.gov/17379808/.
- Scientists including Jennifer Doudna and Emmanuelle Charpentier show that CRISPR can be used as a tool: Martin Jinek et al., 'A programmable dual-RNA-guided DNA endonuclease in adaptive bacterial immunity', *Science* 337/6096 (2012), pp.

816–21: https://www.ncbi.nlm.nih.gov/pmc/articles/
PMC6286148/.

- The first proof that the system worked on human cells: Prashant Mali et al., 'RNA-guided human genome engineering via Cas9', *Science* 339/6121 (2013), pp. 823–6: https://www.ncbi.nlm.nih.gov/ pmc/articles/PMC3712628/.
- Le Cong et al., 'Multiplex genome engineering using CRISPR/Cas systems', *Science* 339/6121 (2013), pp. 819–23: https://www.ncbi.nlm.nih.gov/ pubmed/23287718.

6 Viruses that attack bacteria are known as phages or, more formally, as bacteriophages. There are a huge variety of phages, functioning in many different ways, with various different types of genetic systems among them. Viruses don't always invade a bacterium's genome. That generally happens only when the virus is not reproducing, but merely lying latent within a bacterial cell (lysogeny). This often has an evolutionary advantage. In what is known as 'lytic growth', the production of new viruses often involves viral DNA bursting out of a bacterium's genome, after which it is copied and used to produce new viral particles. Many viruses are virulent (they cause damage to their host), and their DNA is never incorporated into bacterial DNA.

7 Bacteria have other weapons besides CRISPR to defend themselves against viruses. The restriction modification system, for instance, is far commoner than CRISPR, but there are also bacterial suicide systems, as well as others of which we have only a partial understanding. It's likely that there are also others we haven't yet discovered.

8 We know of many different CRISPR systems, and there are presumably many more yet to be discovered. Jennifer Doudna's lab website gives a clear account of several of them: http://doudnalab.org/research_areas/crispr-systems/.

9 Jennifer Doudna and Samuel Sternberg, *A Crack in Creation: The New Power to Control Evolution* (London: Bodley Head, 2017).

10 One account of the history of genetic modification is Gabriel Rangel, 'From corgis to corn: a brief look at the long history of GMO technology', Science in the News

[website] (9 August 2015): http://sitn.hms.harvard.edu/
flash/2015/from-corgis-to-corn-a-brief-look-at-the-long-
history-of-gmo-technology/.

11   There is plenty of interesting reading on how the printing
press was invented in both China and Korea long before
Gutenberg. For those interested, I recommend Elisabeth
Palermo, 'Who invented the printing press?' Live Science
[website] (25 February 2014): https://www.livescience.
com/43639-who-invented-the-printing-press.html.

12   Sweden's first ever printed book was *Dialogus creaturarum
optime moralizatus*. See https://en.wikipedia.org/wiki/
Dialogus_creaturarum.

13   A scientific account of how these various systems work
may be found in Thomas Gaj et al., 'ZFN, TALEN, and
CRISPR/Cas-based methods for genome engineering',
Trends in Biotechnology 31/7 (July 2013), pp. 397–405:
https://pubmed.ncbi.nlm.nih.gov/23664777/.

6. Unruly Peas

1   It goes without saying that much has been written about
Mendel and his peas. For those keen to find out more,
a good place to start is the section on Mendel in the
Encyclopaedia Britannica, which contains a wealth of
interesting information: https://www.britannica.com/
biography/Gregor-Mendel.

2   Mendel's original publication: Gregor Mendel,
'Versuche über Pflanzenhybriden', in *Verhandlungen des
naturforschenden Vereines in Brünn*, vol. 4 (Brno: Vereines,
1866), pp. 3–47: https://www.biodiversitylibrary.org/
item/188027#page/5/mode/1up. For the English translation,
see Gregor Mendel, 'Experiments in plant hybridization',
trans. C. T. Druery, *Journal of the Royal Horticultural
Society* 26 (1901), pp. 1–32.

3   The fact that Mendel's figures are slightly too good to be
true is known as 'the Mendelian paradox'. The scientist
from whom the quotation is taken was Anthony F. W.
Edwards, but many others have also written about this. One
interesting article is Ira Pilgrim, 'The too-good-to-be-true
paradox and Gregor Mendel', *The Journal of Heredity* 75/6

(1984), pp. 501–2: https://doi.org/10.1093/oxfordjournals.jhered.a109998.

4   An article on how the DNA molecule was discovered: Anne Marie Conlon, 'Rosalind Franklin', *New Scientist*: https://www.newscientist.com/people/rosalind-franklin/.

5   On the website of the BBC's *Science Focus* magazine, Hannah Ashworth attempted to answer the question of how long your DNA is and concluded that there are more than two metres in every cell and a total of about twice the diameter of the solar system in the human body as a whole. See https://www.sciencefocus.com/the-human-body/how-long-is-your-dna/#:~:text=If%20you%20stretched%20the%20DNA,diameter%20of%20the%20Solar%20System.

6   Although the issue goes beyond the scope of this book, it's important to bear in mind that it is not just the sex chromosomes which determine which biological sex a body develops. The development of biological sex is a process encompassing many different stages and factors. It's possible for a body to develop in a way that doesn't match its chromosomes: a body may appear to be male or female without having the usual set of chromosomes. Biological sex exists on a spectrum and isn't something that can be simply divided into two groups. There are many different variants. This has been highlighted by the debate around sportswomen with unusually high testosterone levels.

    Another factor, of course, is that many people are a different sex in practice from the one expressed by their body. This may be because they feel that they don't fit the usual definition of either sex, or because they don't identify with their physical sex. This experience is both more important and more genuine than the sex chromosomes an individual happens to possess.

    Much of interest has been written on this subject. Here are a few articles to serve as an introduction to the topic:

- Claire Ainsworth, 'Sex redefined – the idea of two sexes is simplistic. Biologists now think there is a wider spectrum than that', *Nature* (18 February 2015): https://www.nature.com/news/sex-redefined-1.16943.
- Amanda Montañez, 'Visualizing sex as a spectrum

– infographic reveals the startling complexity of sex determination', *Scientific American* (29 August 2017): https://blogs.scientificamerican.com/sa-visual/visualizing-sex-as-a-spectrum/.

- 'The new science of sex and gender – why the new science of sex and gender matters for everyone', *Scientific American* (1 September 2017): https:///www.scientificamerican.com/article/the-new-science-of-sex-and-gender/.

- Alexandra Kralick, 'We finally understand that gender isn't binary. Sex isn't, either', Slate [website] (13 November 2018): https://slate.com/technology/2018/11/sex-binary-gender-neither-exist.html.

7   The Human Genome Project was a vast undertaking. Full details are provided on the project's website: https://www.genome.gov/human-genome-project.

8   The following article, published 20 years after Dolly's birth, summarizes the state of play and future prospects: Sharon Begley, 'It's been 20 years since Dolly. Where's my clone?', Stat [website] (5 July 2016): https://www.statnews.com/2016/07/05/dolly-cloning-sheep-anniversary/.

9   Zhong Zhong and Hua Hua, the two macaques born in January 2018, are the first to be cloned in the same way as Dolly, using somatic cells (non-reproductive cells, taken from the body of an organism). A monkey named Tetra was born in the United States in 1999 through a cloning technique called 'embryo-splitting', preceded by another born in 1997 using a different cloning technique known as 'nuclear transfer'.

Here is a scientific article about Zhong Zhong and Hua Hua (including references to earlier studies): Zhen Liu et al., 'Cloning of macaque monkeys by somatic cell nuclear transfer', *Cell* 172/4 (8 February 2018), pp. 881–7.e7: https://www.cell.com/cell/fulltext/S0092-8674(18)30057-6.

10   The whole of Bill Clinton's speech and the rest of the announcement can be viewed on the National Human Genome Research Institute's YouTube channel. See 'Human genome announcement at the White House, 2000': https://www.youtube.com/watch?v=slRyGLmt3qc.

11   The following article (among others) explains what happened when humans developed the ability to drink milk: Andrew Curry, 'Why we "got milk"', *Scientific American* (31 July 2013): https://www.scientificamerican.com/article/why-we-got-milk/.

7. Gene Therapy among Gilded Temples

1   Hangzhou Cancer Hospital has a website that provides more information: http://www.hzszlyy.com/.

2   There are several good reports on the hospital and its experimental treatments:
- Rob Stein, 'Doctors in China lead race to treat cancer by editing genes', NPR [website] (21 February 2018): https://www.npr.org/sections/health-shots/2018/02/21/585336506/doctors-in-china-lead-race-to-treat-cancer-by-editing-genes.
- Preetika Rana, Amy Dockser Marcus and Wenxin Fan, 'China, unhampered by rules, races ahead in gene-editing trials', *Wall Street Journal* (21 January 2018): https://www.wsj.com/articles/china-unhampered-by-rules-races-ahead-in-gene-editing-trials-1516562360.

3   In 2018, James P. Allison and Tasuku Honjo were jointly awarded the Nobel Prize in Medicine for their discovery of strategies for treating cancer that involve inhibiting mechanisms which function as a brake on the immune system. This press release includes a summary of their research: https://www.nobelprize.org/uploads/2018/10/press-medicine2018.pdf.

A few years previously, *Science* selected this approach as 'Breakthrough of the Year'. You can read more about the underlying research and its impact in Jennifer Couzin-Frankel, 'Cancer immunotherapy', *Science* 342/6165 (20 December 2013), pp. 1432–3: https://science.sciencemag.org/content/342/6165/1432.

The following article looks at the risk of overestimating the results of immunotherapy: Jocelyn Kaiser, 'Too much of a good thing?', *Science* 359/6382 (23 March

2018), pp. 1346–7: https://science.sciencemag.org/content/359/6382/1346/.

4   Since the development of immunotherapy, scientists have also examined various ways of harnessing gene technology to modify immune cells. Chimeric antigen receptor T-cell (CAR-T) therapy has been developed and CRISPR is being tried out on T cells. Here is some further reading for those wishing to look at this in more depth:

- 'CRISPR, CAR-T, and cancer', International Society for Stem Cell Research [website] (27 October 2017): https://www.isscr.org/professional-resources/news-publicationsss/isscr-news-articles/blog-detail/stem-cells-in-focus/2017/10/27/crispr-car-t-and-cancer.
- Hasan Mollanoori et al., 'CRISPR/Cas9 and CAR-T cells, collaboration of two revolutionary technologies in cancer immunotherapy, an instruction for successful cancer treatment', *Human Immunology* 79/12 (December 2018), pp. 876–82: https://pubmed.ncbi.nlm.nih.gov/30261221/.
- Matthew L. Cooper et al., 'An "off-the-shelf" fratricide-resistant CAR-T for the treatment of T cell hematologic malignancies', *Leukemia* 32 (2018), pp. 1970–83: https://www.nature.com/articles/s41375-018-0065-5.
- Shivani Srivastava and Stanley R. Riddell, 'CAR T cell therapy: challenges to bench-to-bedside efficacy', *Journal of Immunology* 200/2 (15 January 2018), pp. 459–68: https://www.ncbi.nlm.nih.gov/pmc/articles/PMC5957501.

5   The website clinicaltrials.gov shows all the clinical trials registered by the US authorities.

6   Three scientific articles about the application of CRISPR to cancer treatments:

- Ping Fan et al., 'Exposing cancer with CRISPR-Cas9: from genetic identification to clinical therapy', *Translational Cancer Research* 7/3 (June 2018): http://tcr.amegroups.com/article/view/22195/html.
- Volkan I. Sayin and Thales Papagiannakopoulos, 'Application of CRISPR-mediated genome

engineering in cancer research', *Cancer Letters* 387 (28 February 2017), pp. 10–17: https://www.sciencedirect.com/science/article/abs/pii/S0304383516301793.

- Stella Baliou et al., 'CRISPR therapeutic tools for complex genetic disorders and cancer', *International Journal of Oncology* 53/2 (August 2018), pp. 443–68: https://www.ncbi.nlm.nih.gov/pmc/articles/PMC6017271/.

7 Here is an article about how China has taken the lead in applying CRISPR to cancer treatment: Dennis Normile, 'China sprints ahead in CRISPR therapy race', *Science* 358/6359 (6 October 2017), pp. 20–1: https://science.sciencemag.org/content/358/6359/20.

8 Choosing to go abroad for medical treatment in countries with laxer laws is both risky and expensive, and there isn't always any regulation to ensure that treatment actually works. Here are some articles on the subject:

- Nahum Méndez-Sánchez and Misael Uribe, 'Biomedical research in Mexico, a perspective from the private medical sector', *Medicina Universitaria* 14/54 (January 2012), pp. 37–41: https://www.elsevier.es/en-revista-medicina-universitaria-304-articulo-bio-medical-research-in-mexico-perspective-X1665579612234379.
- Guillermo Aquino-Jarquin, 'Tighten up Mexico's regulations on human gene editing', *Nature* (26 February 2019): https://www.nature.com/articles/d41586-019-00666-0.
- Vanessa Nason, 'The dangerous practice of seeking medical treatments abroad', *Vice* [website] (25 October 2016): https://www.vice.com/en_us/article/qbnjw5/the-dangerous-practice-of-seeking-medical-treatments-abroad.

8. Panacea

1 Article in *New Scientist* about Irina Conboy and others: Michael Le Page, 'We're nearly ready to use CRISPR to target far more diseases', *New Scientist* (2 October 2017):

https://www.newscientist.com/article/2149129-were-nearly-ready-to-use-crispr-to-target-far-more-diseases/.

2 The two scientific articles below look at how CRISPR can be used in medical science:

- Kiran Musunuru, 'The hope and hype of CRISPR-Cas9 genome editing. A review', *JAMA Cardiology* 2/8 (August 2017), pp. 914–19: https://jamanetwork.com/journals/jamacardiology/article-abstract/2632329.
- Odatha W. Kotagama et al., 'Era of genomic medicine: a narrative review on CRISPR technology as a potential therapeutic tool for human diseases', *BioMed Research International* 2019 (2019): https://www.hindawi.com/journals/bmri/2019/1369682/abs.

But there are also articles that draw attention to problems, such as F. Baylis, 'Counterpoint: the potential harms of human gene editing using CRISPR-Cas9', *Clinical Chemistry* 64/3 (March 2018), pp. 489–91: http://clinchem.aaccjnls.org/content/64/3/489.abstract.

3 A list of all the known diseases caused by a single gene (which totalled 6,538 in November 2019) can be found on the website OMIM Gene Map Statistics: https://www.omim.org/statistics/geneMap.

Another source of information about such disorders is the NCBI webpage 'Single-gene disorders': https://www.ncbi.nlm.nih.gov/books/NBK132154/.

4 Article about the babies treated for bubble boy syndrome: Heidi Ledford, 'Experimental gene therapy frees "bubble-boy" babies from a life of isolation', *Nature* (17 April 2019): https://www.nature.com/articles/d41586-019-01257-9.

Scientific article about the treatment of the babies with bubble boy syndrome: Ewelina Mamcarz et al., 'Lentiviral gene therapy combined with low-dose busulfan in infants with SCID-X1', *New England Journal of Medicine* 380 (18 April 2019), pp. 1525–34: https://www.nejm.org/doi/full/10.1056/NEJMoa1815408.

5 Scientific article reviewing a variety of methods for delivering CRISPR to the body: Christopher A. Lino et al., 'Delivering CRISPR: a review of the challenges and

approaches', *Drug Delivery* 25/1 (2018), pp. 1234–57: https://www.tandfonline.com/doi/full/10.1080/10717544.2018.1474964.

6   News article about the first ever attempt to inject CRISPR into the body: Jocelyn Kaiser, 'A human has been injected with gene-editing tools to cure his disabling disease. Here's what you need to know', *Science* (15 November 2017): https://www.sciencemag.org/news/2017/11/human-has-been-injected-gene-editing-tools-cure-his-disabling-disease-here-s-what-you.

     An article in the *New York Post* a few months later indicated that everything seemed to have gone well: 'Scientists see positive results from 1st-ever gene-editing therapy' (6 February 2018): https://nypost.com/2018/02/06/scientists-see-positive-results-from-1st-ever-gene-editing-therapy/.

7   News article about a clinical trial involving the use of CRISPR in an attempt to cure a genetic eye disorder: 'Eye disorder the focus of new clinical trial using highly touted CRISPR technology', American Optometric Association [website] (29 August 2019): https://www.aoa.org/news/clinical-eye-care/eye-disorder-crispr-technology.

8   News article about the possible ways in which CRISPR could be applied to foetuses in the womb: Megan Molteni, 'CRISPR gene editing is coming for the womb', *Wired* (14 April 2019): https://www.wired.com/story/crispr-gene-editing-is-coming-for-the-womb/.

9   Despite improvements in recent years, the statistics on HIV and AIDS are sobering. The figures quoted here are taken from hiv.gov, a US government site: https://www.hiv.gov/hiv-basics/overview/data-and-trends/global-statistics.

10   News article summarizing the situation as regards the hope of curing HIV with the help of CRISPR: Jon Cohen, 'Curing HIV just got more complicated. Can CRISPR help?', *Science* (15 March 2019): https://www.sciencemag.org/news/2019/03/curing-hiv-just-got-more-complicated-can-crispr-help.

11   A scientific article: Jane M. Carlton, 'Evolution of human malaria', *Nature Microbiology* 3 (24 May 2018), pp. 642–3: https://www.nature.com/articles/s41564-018-0170-2.

12   It's fascinating to see how a disease like sickle-cell anaemia

was able to evolve and how it has persisted. The following article gives a good overview of the disease: Pardis C. Sabeti, 'Natural selection: uncovering mechanisms of evolutionary adaptation to infectious disease', Scitable [website] (2008): https://www.nature.com/scitable/topicpage/natural-selection-uncovering-mechanisms-of-evolutionary-adaptation-34539/.

Also: Graham R. Serjeant, 'The natural history of sickle cell disease', Cold Spring Harbor Perspectives in Medicine [website] 3/10 (October 2013): https://www.ncbi.nlm.nih.gov/pmc/articles/PMC3784812/.

The following scientific article shows that the disease is still subject to positive selection because of malaria: Eric Elguero et al., 'Malaria continues to select for sickle cell trait in Central Africa', *PNAS* 112/22 (May 2015), pp. 7051–4: https://www.ncbi.nlm.nih.gov/pmc/articles/PMC4460506.

13 A news article about the various studies into the use of CRISPR to combat sickle-cell anaemia that are already under way or shortly to be launched: Usha Lee McFarling, 'More than a half-dozen teams are pursuing genetic therapies for sickle cell', Stat [website] (21 February 2018): https://www.statnews.com/2018/02/21/sickle-cell-genetic-therapies/.

A report on one of the ongoing studies: Rob Stein, 'A patient hopes gene-editing can help with pain of sickle cell disease', NPR [website] (10 October 2019): https://npr.org/section/health-shots/2019/10/10/766765780/after-a-life-of-painful-sickle-cell-disease-a-patient-hopes-gene-editing-can-hel.

14 A report by the US Food and Drug Administration (FDA) on the problems facing patients with sickle-cell anaemia: 'The voice of the patient' (October 2014): https://www.fda.gov/media/89898/download.

A news article on the difficulties patients are confronted with: Sharon Begley, '"Every time it's a battle": in excruciating pain, sickle cell patients are shunted aside', Stat [website] (18 September 2017): https://www.statnews.com/2017/09/18/sickle-cell-pain-treatment/.

15 The story of the Tuskegee study is appalling. Here is a good account of what happened and the

long-term effects: Vann R. Newkirk II, 'A generation of bad blood', *Atlantic* (17 June 2016): https://www.theatlantic.com/politics/archive/2016/06/tuskegee-study-medical-distrust-research/487439/.

The US Centers for Disease Control and Prevention (CDC) have published a timeline of the Tuskegee study: https://www.cdc.gov/tuskegee/timeline.htm.

16   *The Immortal Life of Henrietta Lacks* by Rebecca Skloot (New York: Crown Publishing Group, 2010) is a very well-written account of the story around the HeLa cells.

17   The goddess Panakeia is mentioned at the beginning of the Hippocratic oath, a set of principles which doctors and dentists swear to uphold in their professional lives. The oath is attributed to Hippocrates, who lived in Greece from approximately 460 to 370 BC. Panakeia and her four sisters represented five aspects of medicine and health: universal health (Panakeia), cleanliness (Hygieia), recuperation (Iaso), healing (Aceso) and splendour, glory and beauty (Aglaea).

9. Avatars

1   In the United States, the Neurofibromatosis Network provides information and resources for sufferers and their families: https://www.nfnetwork.org/.

In the UK, resources for people with neurofibromatosis type 1 and other disorders with related symptoms can be found at https://nervetumours.org.uk/.

2   The neurofibromin 1 (NF1) gene consists of 350,000 pairs of bases. For more information on the gene itself and the effects of different mutations, see the following page from the US National Library of Medicine: https://ghr.nlm.nih.gov/condition/neurofibromatosis-type-1.

3   There's an excellent article about this project: Ed Yong, 'Turning piglets into personalized avatars for sick kids', *Atlantic* (13 December 2017): https://www.theatlantic.com/science/archive/2017/12/turning-piglets-into-personalized-avatars-for-sick-kids/548204.

4   Here's an article about the project from the University of Wisconsin-Madison: Kelly April Tyrrell, 'To help

kids battling a rare disease, scientists forge a genetic link between people and pigs' (19 December 2017): https://news.wisc.edu/nf1/.

10. A Huge Spider's Web

1   More information about non-coding DNA can be found on the US National Library of Medicine's webpage 'What is noncoding DNA?': https://ghr.nlm.nih.gov/primer/basics/noncodingdna.

2   Despite the Human Genome Project, it's still not quite clear exactly how many genes we have. The following news article gives a good account of the difficulties we face in establishing the number of genes in the human genome: Cassandra Willyard, 'New human gene tally reignites debate', *Nature* (19 June 2018): https://www.nature.com/articles/d41586-018-05462-w.

3   This involves a huge amount of sophisticated detail. There are genes that don't code for proteins, but which control other aspects of the process. There are also genes that can produce many different proteins, a phenomenon known as 'alternative splicing'. This article provides more information: Luv Kashyap and Parul Tripathi, 'Alternative splicing – how one gene can make many proteins', *Bioscience Explained* 4 (2008), pp. 1–10: https://www.semanticscholar.org/paper/Alternative-Splicing-How-one-gene-can-make-many-Kashyap-Tripathi/a7a4e0cf7e019b65ef2547c5b07f5538e1f022cb.

4   A scientific article that gives a good account of how more and more traits are proving to be more complex, and affected by many more genes, than scientists previously thought (one example being height in humans): Evan A. Boyle et al., 'An expanded view of complex traits: from polygenic to omnigenic', *Cell* 169/7 (15 June 2017), pp. 1177–86: https://www.sciencedirect.com/science/article/pii/S0092867417306293.

5   The hereditary component in our height can even exceed 80 per cent. Here is a scientific article that examines different populations: Karri Silventoinen et al., 'Heritability of adult body height: a comparative study of twin cohorts in

eight countries', *Twin Research and Human Genetics* 6/5 (October 2003), pp. 399–408: https://www.cambridge.org/core/journals/twin-research-and-human-genetics/article/heritability-of-adult-body-height-a-comparative-study-of-twin-cohorts-in-eight-countries/3EF884AEA534C90F46F95C9FA3944C84.

6   The link between earwax and breast cancer: Yu Toyoda et al., 'Earwax, osmidrosis, and breast cancer: why does one SNP (538G>A) in the human ABC transporter ABCC11 gene determine earwax type?', *FASEB Journal* 23/6 (21 April 2009), pp. 2001–13: https://faseb.onlinelibrary.wiley.com/doi/full/10.1096/fj.09-129098.

7   Large-scale genetic studies of this type are known as 'genome-wide association studies' (GWAS for short). Here is a scientific article that gives an account of the technique and its potential: Hafsa Tahir et al., 'An overview of genome-wide association study for genetics novices: a review', *Advancements in Life Sciences* 6/3 (2019): http://submission.als-journal.com/index.php/ALS/article/view/523.

And here is a news item about the technique and its limitations: Brian Resnick, 'How scientists are learning to predict your future with your genes', Vox [website] (25 August 2018): https://www.vox.com/science-and-health/2018/8/23/17527708/genetics-genome-sequencing-gwas-polygenic-risk-score.

8   Applying AI and machine learning may be the only way to understand our genes. There are a number of good articles on the subject for those interested in learning more:

- Sarah Webb, 'Deep learning for biology', *Nature* (20 February 2018): https://www.nature.com/articles/d41586-018-02174-z.
- Ashley Yeager, 'Could AI make gene editing more accurate?' *Scientist* (1 May 2019): https://www.the-scientist.com/the-literature/could-ai-make-gene-editing-more-accurate-65781.
- Daniel R. Schrider and Andrew D. Kern, 'Supervised machine learning for population genetics: a new paradigm trends in genetics', *Cell* 34/4 (April 2018),

pp. 301–12: https://www.sciencedirect.com/science/article/pii/S0168952517302251.

- Seonwoo Min et al., 'Deep learning in bioinformatics', *Briefings in Bioinformatics* 18/5 (September 2017), pp. 851–69: https://academic.oup.com/bib/article/18/5/851/2562808.
- Eric J. Topol, 'High-performance medicine: the convergence of human and artificial intelligence', *Nature Medicine* 25 (2019), pp. 44–56: https://www.nature.com/articles/s41591-018-0300-7.

9   It's still not absolutely clear which genes govern a person's exact eye colour. However, the scientific article referenced here, which aims to clarify the matter, identifies more than 16: Désirée White and Montserrat Rabago-Smith, 'Genotype-phenotype associations and human eye color', *Journal of Human Genetics* 56 (2011), pp. 5–7: https://www.nature.com/articles/jhg2010126#citeas.

11. A Modern-Day Horoscope

1   Two news articles, covering the suspect's arrest and the legal proceedings a year later:
- Laurel Wamsley, 'After arrest of suspected Golden State Killer, details of his life emerge', NPR [website] (26 April 2018): https://www.npr.org/sections/thetwo-way/2018/04/26/606060349/after-arrest-of-suspected-golden-state-killer-details-of-his-life-emerge.
- Breeanna Hare and Christo Taoushiani, 'What we know about the Golden State Killer case, one year after a suspect was arrested', CNN [website] (24 April 2019): https://edition.cnn.com/2019/04/24/us/golden-state-killer-one-year-later/index.html.

2   A news story about how the police identified the rapist: 'Swede jailed for child rape 24 years ago in groundbreaking court case', The Local [website] (21 May 2019): https://www.thelocal.se/20190521/swede-jailed-for-child-rape-24-years-ago-in-groundbreaking-court-case/.

For those who read Swedish, here's an article about the change in the law: 'Lagändring ger polisen möjlighet att söka efter okända gärningsmäns familj i DNA-register', Dagens Juridik [website] (28 January 2019): https://www.dagensjuridik.se/nyheter/lagandring-ger-polisen-mojlighet-att-soka-efter-okanda-garningsmans-familj-i-dna-register/.

3   A scientific article about how much of the US population can now be identified using existing DNA data: Yaniv Erlich et al., 'Identity inference of genomic data using long-range familial searches', *Science* 362/6415 (9 November 2018), pp. 690–4: https://science.sciencemag.org/content/362/6415/690.

News article in *Science*: Jocelyn Kaiser, 'We will find you: DNA search used to nab Golden State Killer can home in on about 60% of white Americans' (11 October 2018): https://www.sciencemag.org/news/2018/10/we-will-find-you-dna-search-used-nab-golden-state-killer-can-home-about-60-white.

4   The familial DNA search method used to find the Golden State Killer has raised many questions about its use:

- Eric Ortiz, 'Golden State Killer suspect's capture sparks DNA site privacy fears', NBC News [website] (27 April 2018): https://www.nbcnews.com/news/us-news/golden-state-killer-suspect-s-capture-sparks-dna-site-privacy-n869661.
- Russell Brandom, 'The DNA technique that caught the Golden State Killer is more powerful than we thought', The Verge [website] (11 October 2018): https://www.theverge.com/2018/10/11/17964862/family-dna-crime-search-golden-state-killer-forensics.

5   A news article on the rapid growth of the market in home gene test kits: Mark Williams, 'The lucrative rise of DNA testing: "we created the market for what we do"', *Guardian* (25 May 2017): https://www.theguardian.com/small-business-network/2017/may/25/dna-testing-we-created-the-market-for-what-we-do-living-dna-dnafit-geneu.

An article in *MIT Technology Review*: Antonio Regalado, 'More than 26 million people have taken an at-home ancestry test' (11 February 2019): https://www.technologyreview.com/s/612880/more-than-26-million-people-have-taken-an-at-home-ancestry-test/.

6   Scientific article about the population structure of Iceland: J. T. Williams, 'Origin and population structure of the Icelanders', *Human Biology* 65/2 (April 1993), pp. 167–91: https://www.ncbi.nlm.nih.gov/pubmed/8449480.

7   News article about the genetic quest for Niall: Aoife McLysaght, 'The genetic imprint of Niall of the Nine Hostages', *Irish Times* (24 April 2014): https://www.irishtimes.com/news/science/the-genetic-imprint-of-niall-of-the-nine-hostages-1.1771373.

8   Many journalists have produced good, interesting articles and radio programmes about the problems of using gene tests to investigate one's origins, including interviews with many scientists. Here are a couple of suggestions:
- Rose Eveleth, 'We are family' [an episode of the Flash Forward podcast] (5 June 2018): https://www.flashforwardpod.com/2018/06/05/we-are-family/.
- Kristen V. Brown, 'How DNA testing botched my family's heritage, and probably yours, too', Gizmodo [website] (16 January 2018): https://gizmodo.com/how-dna-testing-botched-my-familys-heritage-and-probab-1820932637.

         Journalists have also written about how gene tests can reveal family secrets: Rachel Cooke, 'Dani Shapiro: "Science will bring an end to these family secrets"', *Guardian* (9 June 2019): https://www.theguardian.com/global/2019/jun/09/dani-shapiro-science-will-bring-an-end-to-these-family-secrets-inheritance.

9   Angelina Jolie's op-ed: 'My medical choice' (14 May 2013): https://www.nytimes.com/2013/05/14/opinion/my-medical-choice.html.

10   Mary-Claire King discovered the connection between mutations in BRCA1 and BRAC2. The article referenced here gives a very interesting account of her life and work: Alice Park, 'Lessons from the woman who discovered the BRCA cancer gene', *Time* (2 June 2014): https://time.com/2802156/lessons-from-the-woman-who-discovered-the-brca-cancer-gene/.

11   There are many good articles about problems with gene testing kits for home use. I can recommend the following:
- Amit Katwala, 'The spurious, questionable

science behind DNA testing kits', *Wired* (30 May 2019): https://www.wired.co.uk/article/dna-testing-kits-science.

- Antonio Regalado, 'How to spend $1900 on gene tests without learning a thing', *MIT Technology Review* (18 October 2017): https://www.technologyreview.com/s/609103/how-to-spend-1900-on-gene-tests-without-learning-a-thing/.
- 'Why you should be careful about 23andMe's health test', *New York Times* (1 February 2019): https://www.nytimes.com/interactive/2019/02/01/opinion/23andme-cancer-dna-test-brca.html.

12 The scientific article on the apple juice study: Stephan W. Barth et al., 'Moderate effects of apple juice consumption on obesity-related markers in obese men: impact of diet-gene interaction on body fat content', *European Journal of Nutrition* 51/7 (October 2012), pp. 841–50: https://link.springer.com/article/10.1007%2Fs00394-011-0264-6.

13 The quotations from Veritas Genetics and Helix are taken from Kristen V. Brown, 'The next pseudoscience health craze is all about genetics', Gizmodo [website] (15 February 2017): https://gizmodo.com/the-next-pseudoscience-health-craze-is-all-about-geneti-1792194708.

14 Doctors and scientists are now beginning to call for such tests to be more strictly regulated.

- Hannah Devlin, 'Senior doctors call for crackdown on home genetic testing kits', *Guardian* (21 June 2019): https://www.theguardian.com/science/2019/jul/21/senior-doctors-call-for-crackdown-on-home-genetic-testing-kits.
- 'Genetic tests: Experts urge caution over home testing', BBC News [website] (17 October 2019): https://www.bbc.com/news/health-50069155.

The problem is also summarized in the following scientific article: Rachel Horton et al., 'Direct-to-consumer genetic testing', *British Medical Journal* 367 (2019): https://www.bmj.com/content/367/bmj.l5688.

15 Plenty of dubious attempts are being made to map genes and link them with everything from income to taste in wine. There are many good articles on the subject that fall

slightly outside the scope of this book, but I recommend the following: Megan Molenti, 'Researchers want to link your genes and income – should they?', *Wired* (12 April 2019): https://www.wired.com/story/researchers-want-to-link-your-genes-and-incomeshould-they/.

16   A number of good articles have appeared about how test data is sold on. Here are a few to begin with:

- Karlin Lillington, 'Don't buy online DNA ancestry tests. You are the real product', *Irish Times* (29 November 2018): https://www.irishtimes.com/business/technology/don-t-buy-online-dna-ancestry-tests-you-are-the-real-product-1.3713619.

- 'Who's making money from your DNA?' [article adapted by Philippa Fogarty from a programme on the BBC World Service presented by Manuela Saragosa and produced by Laurence Knight, broadcast on 27 February 2019]: https://www.bbc.com/worklife/article/20190301-how-screening-companies-are-monetising-your-dna.

- Erin Brodwin, 'After you spit into a tube for a DNA test like 23andMe, experts say you shouldn't assume your data will stay private forever', Insider Inc. [website] (16 February 2019): https://www.businessinsider.com/privacy-security-risks-genetic-testing-23andme-ancestry-dna-2019-2?r=US&IR=T.

- Nicole Martin, 'How DNA companies like ancestry and 23andMe are using your genetic data', *Forbes* (5 December 2018): https://www.forbes.com/sites/nicolemartin1/2018/12/05/how-dna-companies-like-ancestry-and-23andme-are-using-your-genetic-data/.

17   Some good articles about gene doping, genetic testing in sport and the possible implications of these phenomena:

- G. Gregory Haff, 'Will the genetic screening of athletes change sport as we know it?', The Conversation [website] (8 September 2019): https://theconversation.com/will-the-genetic-screening-of-athletes-change-sport-as-we-know-it-122781.

- Colin Moran, 'Explainer: what is gene doping – and will any athletes at Rio 2016 have tried it?' The

Conversation [website] (8 August 2016): https://theconversation.com/explainer-what-is-gene-doping-and-will-any-athletes-at-rio-2016-have-tried-it-63230.
- Kristen V. Brown, 'The search for the Olympian gene', Gizmodo [website] (21 February 2018): https://gizmodo.com/the-search-for-the-olympian-gene-1822975337.

18 A very well-written report on China's genetic tests on Uighurs: Sui-Lee Wee, 'China uses DNA to track its people, with the help of American expertise', *New York Times* (21 February 2019): https://www.nytimes.com/2019/02/21/business/china-xinjiang-uighur-dna-thermo-fisher.html?wpisrc=nl_today_world&wpmm=1.

(A quotation from the article: 'From 2016 to 2017, nearly 36 million people took part in it, according to Xinhua, China's official news agency. The authorities collected DNA samples, images of irises and other personal data, according to Uighurs and human rights groups.

In patent applications filed in China in 2013 and 2017, ministry researchers described ways to sort people by ethnicity by screening their genetic makeup. They took genetic material from Uighurs and compared it with DNA from other ethnic groups. In the 2017 filing, researchers explained that their system would help in "inferring the geographical origin from the DNA of suspects at crime scenes".')

- The Human Rights Watch report: 'China: minority region collects DNA from millions' (13 December 2017): https://www.hrw.org/news/2017/12/13/china-minority-region-collects-dna-millions.
- One of many reports on the appalling 're-education camps': Chris Buckey, 'China is detaining Muslims in vast numbers. The goal: "transformation"', *New York Times* (8 September 2018): https://www.nytimes.com/2018/09/08/world/asia/china-uighur-muslim-detention-camp.html.

Here are two books that examine the links between genes, heredity and origins, both written by fine science journalists:
- Carl Zimmer, *She Has Her Mother's Laugh: The Powers, Perversions and Potential of Heredity* (New York: Dutton, 2018).

- Adam Rutherford, *A Brief History of Everyone Who Ever Lived* (London: Weidenfeld & Nicolson, 2016).

Winding up this chapter, I have a couple of final comments and some extra reading for those who find the topic as fascinating as I do and have managed to get this far. I'm delighted you've found your way to this last section!

- One of the problems with linking genome-wide association studies to both diseases and heredity is the highly skewed nature of the population screened. Various scientists are working to include more people and more data from under-represented population groups. There's a good account of these efforts in Giorgia Guglielmi, 'Facing up to injustice in genome science', *Nature* (16 April 2019): https://www.nature.com/articles/d41586-019-01166-x.
- One aspect of genetics and origins I don't cover in this book is the issue of human 'races'. There are good reasons for its omission, as the concept of 'race' lacks any scientific foundation. It is rooted in obsolete ideas about human genetics. To find out more, the following article provides a good starting point: Elizabeth Kolbert, 'There's no scientific basis for race – it's a made-up label', *National Geographic* (12 March 2018): https://www.nationalgeographic.com/magazine/2018/04/race-genetics-science-africa/.

12. A Place You Didn't Think Existed

1   *Gattaca*, written and directed by Andrew Niccol, is well worth watching. For more details, see IMDb: https://www.imdb.com/title/tt0119177/.

2   Jackie Leach Scully is now (2021) professor of bioethics and director of the Disability Innovation Institute at the University of New South Wales (UNSW). She is the author of many important texts, including the following:
- 'Disability and the challenge of genomics', in Sahra Gibbon et al., eds, *Routledge Handbook of Genomics, Health and Society* (London: Routledge, 2018): https://www.crcpress.com/Routledge-Handbook-of-Genomics-Health-and-Society/

Gibbon-Prainsack-Hilgartner-Lamoreaux/p/
bok/9781138211957.

- 'Feminist empirical bioethics', in Jonathan Ives et al., eds, *Empirical Bioethics: Theoretical and Practical Perspectives* (Cambridge: Cambridge University Press, 2017): https://www.cambridge. org/core/books/empirical-bioethics/empirical-bioethicstheoretical-and-practical-perspectives/ AC0084D3368B82C3AA6D15BBBB6602DC.
- 'Choice, chance, and acceptance', in Erik Parens and Josephine Johnston, eds, *Human Flourishing in an Age of Gene Editing* (New York: Oxford University Press, 2019): https://oxford.universitypressscholarship. com/view/10.1093/oso/9780190940362.001.0001/ oso-9780190940362-chapter-11.

3   Much has been written about experimental attempts to create chimeras, creatures that are a mix of different animals, and sometimes of animals and humans. All the relevant references are given here:

- A long article about Juan Carlos Izpisua Belmonte and his research, from which the quotation in the chapter is taken: Erika Hayasaki, 'Better living through Crispr: growing human organs in pigs', *Wired* (9 March 2019): https://www.wired.com/story/ belmonte-crispr-human-animal-hybrid-organs/.
- A scientific article about creating rat–mouse chimeras: Jun Wu et al., 'Interspecies chimerism with mammalian pluripotent stem cells', *Cell* 168/3 (26 January 2017), pp. 473–86: https://www.cell. com/cell/fulltext/S0092-8674(16)31752-4.
- News article on scientists who created embryos with human and pig cells: Erin Blakemore, 'Human–pig hybrid created in the lab – here are the facts', *National Geographic* (26 January 2017): https:// www.nationalgeographic.com/news/2017/01/ human-pig-hybrid-embryo-chimera-organs-health-science/.
- News article about scientists who created embryos with human and sheep cells: Michael Greshko, 'Sheep–human hybrids made in lab – get the facts',

National Geographic (18 February 2018): https://www.nationalgeographic.com/news/2018/02/sheep-human-hybrids-chimeras-crispr-organ-transplant-health-science/.

Chinese scientists have combined stem cells from human embryos with rabbit embryos. *New Scientist* published the following article on this research: Philip Cohen, 'Human–rabbit embryos intensify stem cell debate' (15 August 2003): https://www.newscientist.com/article/dn4060-human-rabbit-embryos-intensify-stem-cell-debate/.

4   For information about the shortage of donor organs, see https://www.organdonor.gov/statistics-stories/statistics.html (US) and https://www.organdonation.nhs.uk/helping-you-to-decide/about-organ-donation/statistics-about-organ-donation/ (UK).

Again, I recommend Susanne Lundin's book *Organs for Sale: An Ethnographic Examination of the International Organ Trade* (Basingstoke: Palgrave Macmillan, 2015).

5   A scientific article outlining historical attempts to transplant animal organs to humans: David K. C. Cooper et al., 'A brief history of xenotransplantation', *International Journal of Surgery* 23 (November 2015), pp. 205–10: https://www.ncbi.nlm.nih.gov/pmc/articles/PMC4684730/.

6   News article about Su Bing's experiment: Antonio Regalado, 'Chinese scientists have put human brain cells in monkeys – and yes, they may be smarter', *MIT Technology Review* (10 April 2019): https://www.technologyreview.com/s/613277/chinese-scientists-have-put-human-brain-genes-in-monkeysand-yes-they-may-be-smarter/.

- Su Bing defends the experiments in which he transferred human genes to monkeys to modify their brains: Ben Westcott and Serenitie Wang, 'Chinese scientists defend implanting human gene into monkeys' brains', CNN [website] (12 April 2019): https://edition.cnn.com/2019/04/12/asia/monkey-brain-human-china-intl/index.html.
- A scientific article examining the ethics of genetically modifying apes to make them more like humans: Marilyn E. Coors et al., 'The ethics of using transgenic non-human primates to study what

makes us human', *Nature Reviews Genetics* 11/9 (September 2010), pp. 658–62: https://www.ncbi. nlm.nih.gov/pubmed/20717156.

7   Many kinds of research into the dream of optimizing the human body are currently under way. In October 2017, the Wilson Center released a report on the state of play and what may be possible in future: 'How to optimize human biology: where genome editing and artificial intelligence collide': https://www.scribd.com/document/361235830/ How-to-Optimize-Human-Biology-Where-Genome-Editing-and-Artificial-Intelligence-Collide#download.

8   There are many ongoing studies into different ways of using CRISPR to heal or treat various diseases. The list of these studies is constantly changing, so references to particular studies are soon outdated. However, the following website gives a good account of the state of play as regards a number of diseases, the people working on them, and the next stage in the research: Meenakshi Prabhune, 'Diseases CRISPR could cure: latest updates on research studies and human trials', Synthego [website] (17 December 2018): https:// www.synthego.com/blog/crispr-cure-diseases.

An article about ongoing efforts to find a cure for motor neurone disease (amyotrophic lateral sclerosis): Luke Dormehl, 'With CRISPR, geneticists have a powerful new weapon in the battle against ALS', Digital Trends [website] (22 December 2017): https://www.digitaltrends. com/cool-tech/crispr-als-treatment/.

There are several studies under way on the use of CRISPR to tackle viral diseases for which no vaccine exists. See Paul Biegler, 'CRISPR hope for vax-resistant disease treatment', *Cosmos* (20 May 2019): https://cosmosmagazine.com/ biology/crispr-hope-for-vax-resistant-disease-treatment.

The following provide information about some rather more ambitious proposals:

- The World Economic Forum has produced a report on 'How to optimize human biology': (10 November 2017): https://www.weforum.org/agenda/2017/11/ how-to-optimize-human-biology.
- David Masci, 'Human enhancement: the scientific and ethical dimensions of striving for perfection',

Pew Research Center [website] (26 July 2016): https://www.pewresearch.org/science/2016/07/26/human-enhancement-the-scientific-and-ethical-dimensions-of-striving-for-perfection/.

- Nick Bilton, 'The "black ball" hypothesis: is gene editing more dangerous than nuclear weapons?', *Vanity Fair* (28 November 2018): https://vanityfair.com/news/2018/11/is-gene-editing-more-dangerous-than-nuclear-weapons.

For a more general philosophical text on the ethics of human enhancement and transhumanism, I can recommend Karim Jebari's doctoral thesis: *Human Enhancement and Technological Uncertainty: Essays on the Promise and Peril of Emerging Technology* (Stockholm: KTH Royal Institute of Technology, 2014).

9   I write about resurrecting extinct animals and using gene drives in my book *The Re-Origin of Species* (London: Scribe, 2018).

10   There's a great article in *The Atlantic* about how Ginkgo is trying to get us to love genetic modification: Sara Zhang, 'Can this company convince you to love GMOs?' (12 February 2018): https://www.theatlantic.com/science/archive/2018/02/ginkgo-bioworks-gmos/552374/.

American scientists are trying to apply CRISPR to yeast to make wine that doesn't leave you with a hangover: Phyllis Picklesimer, '"Jailbreaking" yeast could amp up wine's health benefits, reduce morning-after headaches', ACES News [website] (16 March 2015): https://aces.illinois.edu/news/jailbreaking-yeast-could-amp-wines-health-benefits-reduce-morning-after-headaches.

11   Scientists hope that viruses modified using CRISPR can become a new weapon in the battle against antibiotic-resistant bacteria: Jessica Miley, 'CRISPR edited viruses are the new enemy of superbugs', Interesting Engineering [website] (23 May 2019): https://interestingengineering.com/crispr-edited-viruses-are-the-new-enemy-of-superbugs.

12   The question of how we should use gene technology is largely about which traits we view as positive or potentially advantageous. It can sometimes be useful to look at things from a different angle. For instance, the

article below argues that people with physical disabilities may actually be those best suited to space travel. In outer space, it's much easier to empty an ostomy bag, for example, than to go to the toilet. Rose Eveleth, 'It's time to rethink who's best suited for space travel', *Wired* (27 January 2019): https://www.wired.com/story/its-time-to-rethink-whos-best-suited-for-space-travel/.

Epilogue – Do You Suffer from Freckles?

1   Here's an amusing, light-hearted news article about the genetics behind red hair, with a headline that holds some added appeal as far as I'm concerned: Susannah Cahalan, 'Science shows redheads have genetic superpowers', *New York Post* (2 December 2017): https://nypost.com/2017/12/02/science-shows-redheads-have-genetic-superpowers/.

The following scientific article contains rather more detailed information: Michael D. Morgan, 'Genome-wide study of hair colour in UK Biobank explains most of the SNP heritability', *Nature Communications* 9 (2018): https://www.nature.com/articles/s41467-018-07691-z.

2   Stephen Hawking's motor neurone disease (amyotrophic lateral sclerosis) would have guaranteed that he was never even born in the first place. Although it's harder to say what would have happened to the mathematician Ada Lovelace, the fact that she died young from uterine cancer suggests that she may have carried genes that increased the risk of cancer.

3   Yuval Noah Harari compares the revolution in biology with the revolution in physics in his books *Homo Deus: A Brief History of Tomorrow* (first published in English in 2016) and *21 Lessons for the Twenty-First Century* (first published in English in 2018).

4   There are problems inherent in viewing the body and the brain as a computer, as pointed out by journalist Kirsten V. Brown in the following article: 'The body is not a computer – stop thinking of it as one', Gizmodo [website] (12 May 2017): https://gizmodo.com/the-body-is-not-a-computer-stop-thinking-of-it-as-one-1795067037.

Another aspect of regarding the body as a machine is

the very limiting nature of such a view. Rose Eveleth has written an excellent piece on how wrong-headed such a way of thinking could be: 'Transhumanism is tempting – until you remember Inspector Gadget', *Wired* (27 May 2019): https://www.wired.com/story/what-inspector-gadget-can-teach-us-about-transhumanism/.

5   If you've got this far, you deserve a cake! Here's a recipe for a delicious, chocolatey Sachertorte: https://www.seriouseats.com/recipes/2012/01/sachertorte-viennese-chocolate-cake-apricot-recipe.html.